D1273153

Glass and Archaeology

Studies in Archaeological Science

Consulting editor G. W. DIMBLEBY

Other titles in the Series

The Study of Animal Bones from Archaeological Sites
R. E. CHAPLIN

Land Snails in Archaeology
J. G. EVANS

Methods of Physical Examination in Archaeology
M. S. TITE

Ancient Skins, Parchments and Leathers
R. REED

Soil Science and Archaeology
S. LIMBREY

Fish Remains in Archaeology
R. W. CASTEEL

Principles of Archaeological Stratigraphy
E. C. HARRIS

Animal Diseases in Archaeology
J. BAKER and D. BROTHWELL

Prehistoric Mining and Allied Industries
R. SHEPHERD

Glass and Archaeology
S. FRANK

Glass and Archaeology

Susan Frank
The Library, University of Sheffield

1982

ACADEMIC PRESS
A Subsidiary of Harcourt Brace Jovanovich, Publishers
London New York
Paris San Diego San Francisco São Paulo
Sydney Tokyo Toronto

ACADEMIC PRESS INC. (LONDON) LTD.
24/28 Oval Road
London NW1

United States Edition published by
ACADEMIC PRESS INC.
111 Fifth Avenue
New York, New York 10003

British Library Cataloguing in Publication Data

Frank, Susan
Glass and archaeology.
1. Glass 2. Archaeology
I. Title
666′.1′02493

ISBN 0–12–265620–2

LCCN 810729

Photoset by Paston Press, Norwich
Printed in Great Britain by St Edmundsbury Press

Preface

Archaeology is defined in the Concise Oxford Dictionary as the study of antiquities, especially of the prehistoric period. Many would quarrel with this definition: as one with an interest in industrial archaeology I find it narrow and rather old-fashioned. However, archaeology does involve study and the studies nowadays are much more broadly based than even a few years ago. The latest scientific methods are aplied to the investigation of materials, whilst developments in economic, social, historical and artistic fields all play a part in building up a more complete picture. Thus the people who are involved in archaeological studies come from a wide variety of subject backgrounds and if they are studying a particular type of material they need to look at it from many different aspects.

I have tried in this book to build up a comprehensive guide to glass which will be of interest to all those who study it. In this respect it may be said that I have leaned towards a scientific rather than an artistic or social interpretation. There are two reasons for this. Firstly, there are many good books on the artistic and historical aspects of glass which are freely available. However, for the archaeologist whose work involves the scientific study of glass, there is less choice. Most books have been produced for practising glass technologists and are often empirical in nature. The nature of glass is such that it is much more difficult to understand in a scientific way than most other materials, and this is shown by the fact that advances in glass science tend to be reported only in the specialist literature. Secondly, having been trained as a scientist and having developed into a person with more general interests, I believe that a basic understanding of the nature of glass, in all its aspects, is vital to those whose work involves them in the study of this fascinating material.

The book is a handbook on glass for archaeologists. The person who studies glass must, in the course of his or her investigations, use methods that are common to archaeological studies in general and it is not the aim of this book to describe these techniques. Thus, for example, methods of sampling remains from sites will not be considered, as they can be found elsewhere in standard texts.

As a book of this type can only be an introduction, I have considered it very important to show people how they can find out things for themselves. I have never found the practice of providing a long list of unannotated references particularly helpful in this situation, so I have restricted myself to citing and describing key references in the text, and providing an annotated bibliography for further reading at the end of a chapter, as appropriate. My final chapter is a guide to information retrieval in glass studies. References can become dated, and therefore valueless, but once correct methods of information gathering have been mastered the student or research worker can keep abreast of the latest developments in his or her field of study.

January 1982 Susan Frank

Acknowledgements

I should like to acknowledge the help of the following people in the preparation of this book: Professor R. G. Newton for his kind advice and assistance: Mr E. S. Wood for his information on Wealden glassmaking sites; Mr P. V. Addyman and his colleagues at the York Archaeological Trust for information on the Coppergate excavations, York; Dr J. O. Isard for discussions on glass durability; Mr D. Ashurst and Mr D. Crossley for the use of their site plans; Mrs S. Saxby and Mr R. Wilson for photographic work; and Mrs J. Holmes, Librarian of the Joint Library of Glass Technology, University of Sheffield.

Contents

1

Glass—its Structure and Properties

Introduction

Materials of concern to archaeologists are often relics of the past which have been preserved. They are rarely found in their original state, their condition having been altered by many external agents over a long period of time. Their state of preservation also depends upon their own nature and form. These factors may combine in surprising ways: thus we can still examine pieces of delicate lace netting found in Egyptian tombs, whilst decaying nineteenth-century structures of brick and iron exposed to weathering in pollution-laden atmospheres can present problems of interpretation to the industrial archaeologist.

Glass is no exception as a material of archaeological interest. The state in which it is found depends upon its original structure and composition, and upon the agents which have acted upon it since it was made. By glass, we usually understand a hard material, often transparent or translucent, which is made by heating together a mixture of materials such as sand, limestone and soda at a very high temperature to form a liquid. When this liquid is taken from the furnace it stiffens rapidly as it cools until at about 500°C it resembles the glass we see in windows or formed into bottles. The major ingredients of the commonest commercial glasses are sand, limestone and soda: the soda acts as a flux and the limestone as a stabilizing agent to give a durable glass. Ancient glasses are also frequently of this soda–lime–silica composition: the other major type that is found is a potash–lime–silica glass, where potash replaces the soda as the fluxing agent. However, before we can deal with the complexities of ancient glass compositions and interpret glass remains, we must start by understanding something of the basic physical and chemical nature of glass.

The nature of the glassy state

The solids that we come across in everyday life are usually composed of numerous tiny crystals. A crystal is a substance which has solidified from the liquid into a definite geometrical form, reflecting the arrangement of the constituent atoms. These are packed together in a perfectly regular manner to form a repeating network or lattice. Large single crystals can be found in nature or produced in the laboratory, but a piece of iron of the type found during excavations is polycrystalline, being composed of millions of interlocking crystals all pointing in different directions.

In contrast, glass is a material which has never crystallized, and has become rigid whilst still retaining its liquid structure. This is a departure from the usual pattern of events: when most liquids are cooled a sudden change occurs in their structure at a particular temperature and they freeze to form crystals. In a liquid the atoms are joined to one another but not in any regular extended three-dimensional pattern: they form a random structure. The heat energy contained in the system manifests itself in two ways. The atoms vibrate, and motion of complex atoms or molecules from one position to another also occurs. Some of the interatomic bonds are constantly being broken and reformed so that the liquid can flow. As the temperature is lowered heat energy is removed from the system and at a certain temperature, the freezing point, a discontinuous change takes place. The liquid becomes a solid, the structure changes from random disorder to regular order and the material no longer flows: below the freezing temperature the heat energy of the material is manifest as vibrations of the atoms about their fixed positions in the crystal.

However, as we cool down the liquid that is to become a glass from very high temperatures to room temperature we see no discontinuous changes taking place. It simply gets stiffer and stiffer until it is rigid and effectively a "solid", but still with the internal structure of a liquid. Because it has been cooled far below the temperature at which thermodynamic considerations indicate freezing should take place it is known as a "supercooled" liquid.

A supercooled liquid is in what we call a "metastable" state. Generally speaking, if several states are possible for a system the most stable state is the one in which the energy of the system is least: thus stones fall to the ground when dropped and water runs downhill. At the freezing point an ordinary liquid changes to a solid because in this way the internal energy is minimized and the system is stable. The energy of the system is appropriate to the temperature of the system. A supercooled liquid, however, contains more heat energy than is appropriate for its temperature and it is thus unstable, but it may only be able to reach the stable (crystalline) state by passing through an intermediate state of higher energy. Consider a book standing alone on a shelf. By falling over it can reach a stable state and release energy, but it needs extra energy in the form of a push to reach this state: it is said to be in metastable equilibrium. Similarly the glass-forming liquid requires an energy input in order to attain the crystalline state and this arises because of the peculiar nature of its chemical bonds.

A liquid which may form a glass shows a very rapid increase in stiffness, or viscosity as the temperature drops below that at which the glass is "melted", typically 1350–1600°C. The unit of viscosity, a measure of the degree of stiffness or resistance to flow, is the poise, and it is perhaps easiest to understand what this means by looking at some typical values of viscosity for substances which we come across in everyday life and comparing them with values for glass at various temperatures. At the "melting" temperature the viscosity of glass is about 100 (10^2) poises, comparable with the value of about 200 poises for glycerin at room temperature. By comparison water has a viscosity of about 0·01 poises at room temperature, and light machine oil a viscosity about one poise. Just below the melting temperature is the most dangerous region for crystallization and the glass must be cooled very quickly through this range, giving crystals no chance to form and grow. The glasses are worked (to form containers, window glass, etc.) at lower temperatures (typically 600–700°C for commercial glasses) where the viscosity is perhaps 100 to 10 000 times greater and by the time that the glass reaches room temperature the viscosity is very high, about 10^{20} poises. The reason why the glass-forming liquid can remain in the metastable state now becomes apparent. The viscosity is a measure of the ease with which the glass can flow, and this in turn depends upon the strength of the intermolecular bonds. High viscosity implies high bond strength. The way in which the glass-forming liquid would pass from the metastable state to the stable, crystalline state is by breaking the intermolecular bonds in the liquid and making new bonds to form a regular crystal lattice. However, the original bonds are too strong for this to happen: not enough energy is available to break them and the glass-forming liquid fails to crystallize. This difficulty increases as the temperature falls and by the time that the glass reaches room temperature it may be said that it is a liquid which has become too cold to freeze.

Figure 1 shows how a given property, the volume, varies with temperature for a substance that can exist as a solid, liquid, and glass.

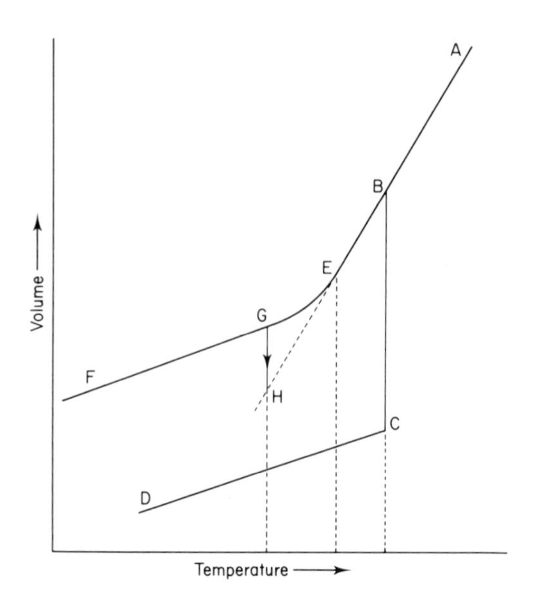

Fig. 1. The relationships between the glassy, liquid and solid states.

Starting with the liquid at A, we cool it. Freezing would be expected to occur at a temperature corresponding to point B. From A to B the liquid contracts by two means. The amplitude of the vibratory motion of the atoms decreases as the temperature falls and this has the effect of reducing the interatomic spacing and so causing the material to contract slightly. As well as this normal thermal contraction there is configurational shrinkage. Rearrangement of the interatomic bonds occurs as the temperature is reduced, in order to reach a stable configuration at any particular temperature and this causes the material to assume a less open structure which occupies less space. If the material crystallizes at B there will be a sudden decrease in volume, followed by thermal contraction only from C to D (room temperature). The crystal undergoes no configurational change. If crystallization is avoided at B, the supercooled liquid continues to shrink from B to E by thermal contraction and configurational shrinkage. At E, however, the slope of the volume-temperature curve decreases and from E to F the substance, although still retaining its liquid structure, shrinks only by thermal contraction. (The volume is larger than for the crystalline state, corresponding to a more open structure in the glass.) The temperature at which this occurs is known as the transformation temperature. Actually, this temperature is not sharply defined but covers a range of about 50°C, the transformation range. As we saw previously it becomes increasingly difficult for configurational changes to occur as the temperature is reduced and when the transformation temperature is reached the glass is so viscous that configurational adjustments do not have time to occur: they cannot keep pace with the rate at which the temperature is falling. The transformation temperature varies widely for different substances, but the viscosity at that temperature is approximately the same for all glasses, 10^{13} poises. For this reason, 10^{13} poises, the viscosity at which configurational adjustments cease under normal conditions, is taken as the dividing line between supercooled liquids and glasses.

If the temperature of the glass is reduced to the value corresponding to point G on the curve, and then held there for a sufficient length of time, the configurational adjustments appropriate to the temperature have time to take place and the volume decreases to the value corresponding to point H. This process is known as stabilization. It is impossible to follow the line EH very far in practice because of the longer and longer times necessary. We can say that in the transformation range the time-scale of the experimental measurements "crosses over" the time-scale of the atomic adjustments.

Related to the stabilization process is the effect of the cooling rate on the final glass volume. Figure 2 shows how the transformation temperature varies with cooling rate.

A rapidly cooled glass follows line 2. If it is cooled more slowly the configurational shrinkage can keep pace with the cooling to a lower temperature (line 1) and the final volume occupied by the glass at room temperature is smaller. Thus the rate of cooling of the glass affects its final internal structure. The faster the glass is cooled the higher the temperature at which configurational rearrangement effectively ceases. This temperature is known as the fictive, or configurational

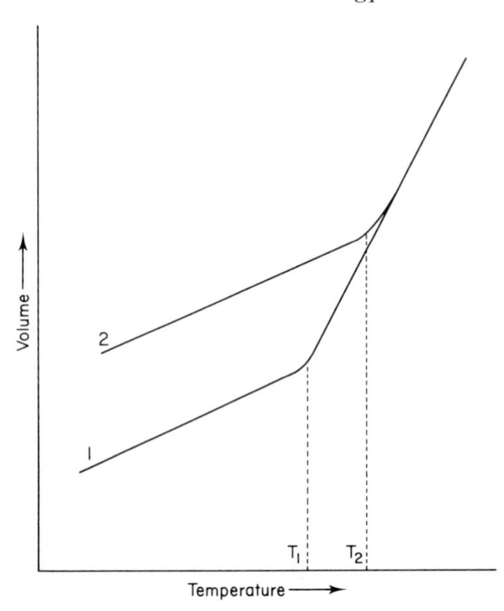

Fig. 2. Variation of final glass volume with cooling rate. 1. Slow cooling. 2. Fast cooling.

temperature of the particular glass sample. A glass has a given value of fictive temperature if its configuration corresponds to that which would be in equilibrium at the temperature named. In Fig. 2 the fictive temperatures of the two samples with fast and slow cooling are T_2 and T_1 respectively, typically around 500°C for ordinary glass, which at room temperature is very slowly shrinking, a contraction which will continue until the fictive temperature reaches the actual temperature. This process is so slow, however, that many millions of years would pass before this happens. Speaking generally, glass can be thought of as a "slow-motion" liquid, having flow properties which are similar, on a vastly increased time scale, to those of ordinary liquids.

We have now explained briefly the characteristics of the glassy state but before we can discuss the nature of glass remains it is necessary to examine the types of material which can form glass.

The structure of glass

Glasses have liquid-like structures, and most are solutions, i.e. homogeneous mixtures of substances of dissimilar molecular structure. There are various types of solutions that we come across in everyday life, such as solids in liquids or liquids in liquids. If a solid is dissolved in a liquid we can prepare solutions containing widely differing amounts of the solid: in fact we are preparing systems with a wide range of compositions. This is in contrast to chemical compounds, where the constituent parts combine in fixed amounts, giving a much more limited composition range. Because they are solutions rather than exact chemical compounds glasses can be formed with a great variety of compositions and properties, which explains their usefulness for a wide range of industrial, research, and domestic applications. In later chapters we shall see that the old glassmakers often had little idea of what they were putting into their glasses: the fact that they generally produced usable materials is a reflection of this useful property of glass.

The major constituent of most common glass is silicon dioxide (silica), SiO_2, derived from ordinary sand. In its crystalline form its basic structure is that of a tetrahedron, with four oxygen atoms surrounding a cental silicon atom (Fig. 3).

This structure is dictated by the nature of the interatomic bonds and the relative sizes of the constituent atoms (the silicon is much smaller than the oxygens). Each oxygen atom is shared between two silicon atoms: the structure is built up by the sharing of corners by pairs of the "oxygen tetrahedra". Many complex, open, three-dimensional structures can be built up in this way, to give various forms of crystalline

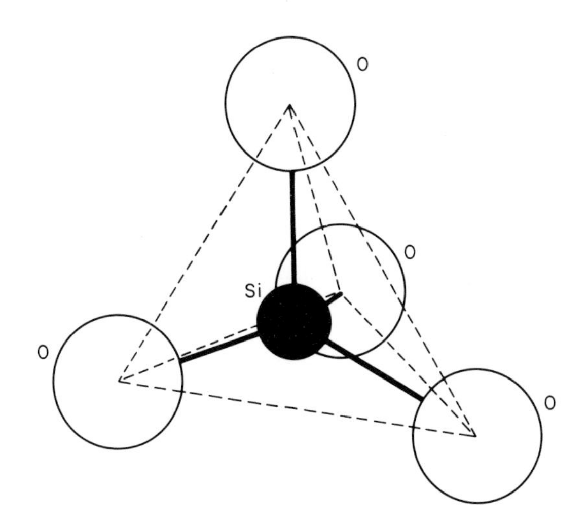

Fig. 3. The basic structural unit of crystalline silica.

silica. The four bonds from the silicon atom to the surrounding oxygens are strongly directional and tend always to keep the same relative orientations, but the silicon–oxygen–silicon angle between the two tetrahedra can vary and tetrahedra can rotate relative to one another, the shared oxygen being the "pivot point". These factors are all important in explaining the role of silica as a glass former.

As can be seen from Fig. 3, the single structural crystal unit for silica is quite difficult to represent on a flat sheet of paper: very little can be demonstrated in two dimensions when we consider the overall crystalline structure and the even more complex structure of the glass. However, we may demonstrate the essential principles by considering possible structures for an imaginary two-dimensional oxide A_2O_3. The oxygen triangles with the atom of the imaginary element A at the centre replace the units of the real system, SiO_2 tetrahedra with oxygen atoms at the four corners. This approach was first used in the early 1930s when a random network model was proposed for the structure of glassy silica. The random network theory remains the basis of much modern research on the structure of the glassy state.

Figure 4 shows the structure of the oxide in a crystalline form, and as a two-dimensional glass. The third structure is a mixed glass formed from the same oxide plus the oxide of a metal, such as sodium, which forms singly charged ions.

In both the crystal and the glass corners are shared between oxygen triangles, but in the glass the triangles are arranged irregularly and the structure is more open than in the crystalline form. It can be seen from Fig. 4(b) that variation in the angle between the two bonds from the oxygen atoms (the two-dimensional analogue of the angle variation and rotation that occurs in real glasses) plays an essential part in allowing the formation of the random, glassy structure. However, the structure is tightly braced because the bonds between the atom A and the three oxygen atoms surrounding it are strongly directional. In the real case of silica glass these strong silicon-oxygen bonds form a tightly braced structure in three dimensions. This explains why pure silica glass, "fused silica", does not soften until very high temperatures are reached, as a great deal of energy is required to disrupt the structural arrangement. A temperature of over 1700°C is necessary to melt quartz sand, and even so the molten material is quite viscous. By the time the glass reaches 1300°C its viscosity has already risen to 10^{12} poises, far above the usual working viscosity for glass of 10^3 to 10^6 poises. The addition of metallic oxides such as sodium oxide (Na_2O)

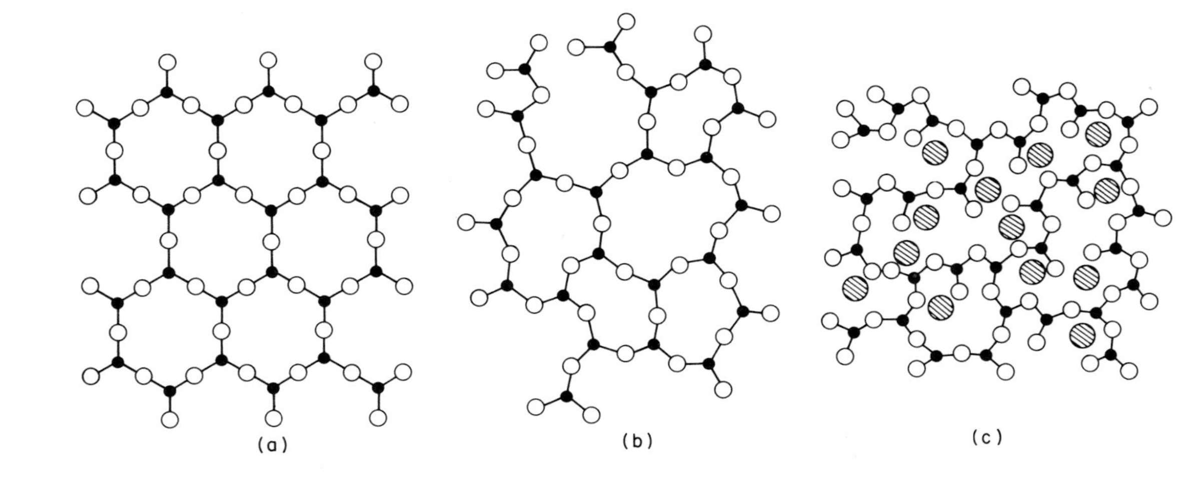

Fig. 4. Structure of an imaginary two-dimensional oxide, A_2O_3. (a) in a crystalline form (b) as a glass (c) as a glass modified by the addition of singly charged metallic ions. (Open circles represent atoms of O and filled circles atoms of A. Shaded circles represent atoms of the singly charged metallic ion M.)

or calcium oxide (CaO) lowers the viscosity so that the required temperature for working a commercial soda–lime–silica glass is only 600–700°C. Why this should happen is explained by reference to the two-dimensional analogue shown in Fig. 4(c). The positively charged metallic ion, M, introduced as an oxide M_2O is located within the spaces in the open glassy structure. Thus the structure is more tightly packed with atoms and it might seem that the glass should become even more viscous. However, we have now introduced extra oxygen atoms and it is no longer possible for every oxygen to be bonded to two atoms of A. The oxygens which are bonded to only one atom of A are called "non-bridging" oxygens. For each extra oxygen atom introduced one of the bonds in the network is broken and two non-bridging oxygens are produced. These have a negative charge. As it is necessary to preserve electrical neutrality within the structure, the positively charged metallic ions, M, occupy spaces in the network near to non-bridging oxygens. The continuity of the network is broken up by the formation of non-bridging oxygens and the new bonds, between them and the metallic ions, are weaker and essentially non-directional. The result is that the structure is less well braced and at a given high temperature the viscosity of a glass containing suitable metallic ions is much less than that of pure silica. This modified glass can be melted and worked at much lower temperatures and thus the production of glass articles of all types becomes a practical proposition.

Although glasses can be formed from, and modified by, a wide variety of oxides, silica, SiO_2, is by far the commonest glass former and is the basis of glasses of interest to archaeologists. Common oxides which modify the structure, disrupting the continuity of the network, include those of sodium, calcium, magnesium and potassium. Lead oxide, which has been of considerable importance in the history of glassmaking, also has the effect of lowering the viscosity. Of course, oxides are not added simply as fluxing agents to lower the melting point: for example lime (CaO) and magnesia (MgO) are added to improve the chemical resistance and many desired physical and chemical properties are now obtained by altering the glass composition.

Properties of glass

Although many of the properties of glass arise as a result of their composition, we shall discuss here those properties which are common to most glasses and which are direct consequences of the physical and chemical nature of the glassy state.

Transparency

We usually think of a glass as being transparent, though of course it may be coloured. Apart from certain single crystals, solids are not generally transparent: transparency is much more characteristic of the liquid than the solid state.

In a crystalline solid, which is usually made up of many tiny crystals, light in its passage through the solid is reflected at each internal boundary. Some light is lost at each reflection, with the result that the material is effectively opaque. A liquid or a glass, however, is structurally a large molecule, containing no internal surfaces or discontinuities having any dimensions approaching the wavelength of visible light. The light can pass through the glass virtually unhindered and thus the glass is transparent. We may see why this happens by considering what happens to the waves rolling in on a beach. The pebbles on the beach (individual units in the glass structure) are very much smaller than the dimensions of the wave (visible light) and have no effect on its passage. However, when the waves encounter rocks of the same dimensions as themselves (internal grain boundaries) they are broken and scattered and can no longer proceed in their original direction.

We are talking here about visible light. Common glass is in fact opaque at infrared and ultraviolet wavelengths, because at these wavelengths the frequencies of the incident radiation are in resonance with the frequencies of molecular vibrations within the glass and this results in its absorption by the glass. If impurities are present in the glass this can also result in absorption at visible wavelengths: very small amounts of impurity can lead to glasses that are tinted or almost opaque. This is the reason for the greenness, or even apparent blackness, of much old glass, which is caused by the presence of small amounts of iron oxide. The ancient glassmakers had no means of getting rid of this impurity in their raw materials, and even to-day, with strict quality control of materials going into the melt, iron oxide is still a problem. The colour that it causes may be seen by looking edgeways on at a piece of apparently colourless modern window glass. Of course, the glassmakers also added colouring agents deliberately, but even so the power of some of these agents meant that they had to use special techniques, for example when producing red glass for use in windows (see Chapter 2).

Resistance to corrosion

Many glasses are very durable, i.e. extremely resistant to corrosion, and their composition can be chosen to enhance this property. However, ordinary glass when placed in contact with water (as may be the case at an archaeological site or in a damp atmosphere) can suffer corrosive attack, because the water leaches out the alkali, sodium or potassium, from the surface. Unfortunately the situation is extremely complex and little is understood about the basic mechanisms involved, certainly in the case of complex ancient glasses. Durability of glass has been studied extensively for many years, and in recent years a whole range of sophisticated scientific techniques have been applied to the study of the leached glass surface and the solutions produced. These have given some insight into the factors involved in glass decomposition but the actual mechanisms continue to be a matter of considerable discussion. If an ordinary glass is placed in contact with an aqueous solution, alkali ions are extracted from the glass in preference to silica and therefore an alkali-deficient layer is formed on the glass surface. The formation of this layer usually reduces the rate of alkali extraction by forming a barrier through which further alkali ions must diffuse before they can be brought into solution. However, as alkali passes into solution the pH value of the solution increases which in turn increases the attack on the silica network. (The pH value is a measure of acidity/alkalinity of a solution. The higher the pH value the higher the alkalinity and the greater the rate of attack on a glass.) Dissolution of the network occurs, so reducing the thickness of the leached surface layer and favouring an increase in extraction rate of alkali. So the attack progresses.

Many factors affect the rate of attack of a glass. The quantity of alkali extracted from a glass in a given period of time increases with increasing temperature, so temperature variations would obviously affect a specimen, but considering the fluctuations which must have occurred for most objects it is unlikely that any valid conclusions could be drawn for a particular piece. It has also been shown that if the leaching solution is renewed, rather than being allowed to remain in contact with the surface, the amount of silica extracted (a measure of the break-up of the glass network) decreases. (This is to be expected because the pH value of the attacking solution does not have a chance to rise so high if the solution is constantly being renewed.) This observation would obviously have implications for archaeological specimens buried in contact, say, with moist soil or subject to running water in river or marine environments, but it is only possible at present to make statements about relatively simple glasses observed under test conditions where other factors are carefully controlled. Marine specimens in any case present extra problems because of the special environment in which they are found, often at some depth and buried in sediment of complex composition. An interesting example is provided by heavily weathered glass retrieved from the harbour at Port Royal, Jamaica, about which there has been considerable, but not conclusive discussion. The sediment from which the glass was retrieved was dark blue: such sediments tend to be reducing in nature and to contain a large amount of marine humus. It is also likely that the glass was subject to bacterial action, since bacterial concentration tends to be high near the shore. Both these factors would cause a reduction in pH values and the environment of the glass would be acidic. In this situation the glass would be attacked (in a different way from attack in an alkaline environment but still producing a silica-rich surface layer) and would show the observed heavily weathered surface.

Other factors which affect the rate of alkali extraction from glass are the nature of the surface and the surface area exposed, the nature of the leaching solution, and, very importantly, the composition of the glass. In general the rate seems to decrease with decreasing alkali content of the glass and with decreasing ionic radius of the alkali ion, also when part of the silica is replaced by almost any divalent oxide. Potash glass (or, more strictly speaking, glass where potash is the major alkali) is in general much less durable than soda glass, and this has important consequences for archaeology because medieval glass is usually a potash glass which is scarce in comparison with Roman glass: the Romans produced a relatively stable glass containing soda as the major alkali. (The effect is not as straightforward as this, as the other constituents of ancient glasses such as calcium also play an important part.) The chapters on the history of glassmaking, conservation and ancient glass compositions go into this matter in greater detail. It may be said here that observations on the durability of ancient glasses are almost entirely empirical in nature, and necessarily so, because the knowledge to interpret simple systems is only just being developed: the complexities of old glasses make them vastly more difficult to understand.

In practical terms, glasses undergoing corrosion deteriorate, often developing a network of fine surface cracks, a phenomenon known to the old glassmakers as crizelling. This was a considerable problem, as the balance of constituents was at that time very much a matter of trial and error, and often too much fluxing agent was added in an attempt to bring down the melting temperature, resulting in a glass highly susceptible to attack.

The beautiful rainbow colours that develop on the surface of some old glass are caused by interference effects of light in very thin layers that have developed as a result of a particular type of corrosive attack. Such glass is usually very fragile and must be treated with extreme care: otherwise the surface flakes off and the glass can shatter. These colours should not be taken as conclusive evidence of age: errors in compositional balance still occur and fakers can duplicate the effect without much difficulty.

Brittleness

Glass when newly formed, with a perfect surface, is very strong, about five times as strong as steel. This may seem strange, but theoretically glass should be very strong because of the nature of its interatomic bonds. In practice the strength is very much less than the theoretical value. One of the main causes of this loss of strength is the presence of surface defects, such as those caused by chemical corrosion or mechanical abrasion. These flaws can be very small but because glass is rigid they act to concentrate any applied stress over only a few interatomic bonds at the apex of the crack. Under these conditions the strong bonds break and fracture occurs. Once started it has a high probability of spreading right across the material because there are no internal grain boundaries to stop it. Thus a piece of glass will often shatter suddenly when subject to a stress.

This concept of stress is also important in cooling the glass during the manufacturing process. If the glass is cooled too rapidly it does not have time to release stresses set up within it during cooling: these are "frozen-in" and can cause the glass to shatter when it becomes a solid. This may occur spontaneously, or when a tiny flaw is produced on the surface. There is no need for applied force in this case as the stresses are built-in and the glass will shatter. In order to avoid these internal stresses, glass articles are subjected to a controlled heat treatment after manufacture, a process known as annealing. The temperature is raised to that which will allow internal stresses to be relaxed by flow within the glass (but not so high that the article will deform) and held there for an appropriate time. It is then slowly reduced to a point well below the transformation range and afterwards more rapidly to room temperature.

Bibliography

Advances in glass science are reported in journal articles, conference proceedings, research reports, and other forms of literature, and can be traced using the publications mentioned in Chapter 8. However, for the reader who wants a non-specialist account there are very few modern texts which provide a clear, straightforward explanation of this complex subject. The following books and review articles are ones which I have found helpful in this respect.

Holloway, D. G. (1973). *The physical properties of glass.* (Wykeham)

The first chapter of this book, on the composition and structure of glass, gives a good explanation of the physical chemistry of the glassy state.

Jones, G. O. (1971). *Glass.* 2nd ed. revised by S. Parke. (Chapman and Hall)

Although this book might prove difficult for the non-scientist it gives a comprehensive and systematic account of the thermodynamics, structure and properties of the glassy state.

Maloney, F. J. T. (1967). *Glass in the modern world: a study in materials development.* (Aldus)

Chapter 1: Glass is a liquid, and Chapter 2: The properties of glass, give a very clear and straightforward explanation of glass structure and properties. The whole book is well worth reading and is probably the best text for the non-specialist who is interested in the scientific and technological aspects of glass. It is attractively produced, easy to read, with good photographs and diagrams.

Rawson, H. (1980). *Properties and applications of glass.* (Elsevier)

A reference book providing an up-to-date account of glass science.

Stanworth, J. E. (1976). Some aspects of the development of glass technology over 60 years. *Glass Technology,* **17**(5), 194–204.

Part of this review article surveys the literature on the structure and constitution of glass, with reference to 50 original papers.

Bamford, C. R. (1977). *Colour generation and control in glass.* (Elsevier)

For those who want to study this important aspect of glass in detail there is a description of the physical basis of colour in glass and the effect of particular elements.

Biek, L. and Bayley, J. (1980). Glass and other vitreous materials. *World Archaeology*, **11**(1), 1–25.

This is a very useful review, with over one hundred references, which discusses recent analytical results and their interpretation, particularly in the fields of coloured and lead glasses. Stress is laid on melting times and temperatures and furnace atmospheres, as well as on compositions. Current work on newly excavated material from English sites is described.

Newton, R. G. (1980). Recent views on ancient glasses. *Glass Technology*, **21**(4), 173–183.

A review, containing nearly one hundred references, discussing developments in knowledge which have taken place during the last ten years from a technological point of view. This article gives a very good overview of recent research.

Weier, L. E. (1973). The deterioration of inorganic materials under the sea. *University of London. Institute of Archaeology. Bulletin*, **11**, 131–163.

A comprehensive and detailed review article of general interest to the marine archaeologist because it covers physical and chemical aspects in detail, together with a discussion of stone, pottery and metals as well as glass.

Paul, A. (1977). Chemical durability of glasses: a thermodynamic approach. *Journal of Materials Science*, **12**, 2246–2268.

Although written from a particular viewpoint, this paper is a review of work on glass durability and is worth reading as an overview of a complex subject.

2

The History of Glassmaking

Introduction

Glass has been made in a variety of forms and ways from very early times and in many areas of the world. The following survey can only describe the major developments, and for more detailed accounts you should consult texts which deal with specific aspects, a selection of which is given in the bibliography to this chapter.

The first glassmakers

Glass used as a glaze predates its use as an independent substance. Glazed steatite and faience objects were made in northern Mesopotamia in the fifth millenium BC, and were exported to other parts of the ancient East including Egypt where the art of glazing objects may have been introduced towards the end of the fourth millenium BC. A few glass beads have been found in VIth Dynasty Egyptian tombs, but glass vessels, formed on a removable core, appear suddenly around 1500 BC in northern Mesopotamia. Moulded objects were also made and exported from here to Asia Minor, Syria, Palestine, Cyprus, Egypt, and the Mycenaean sites in mainland Greece, probably leading to the establishment there of a glass-moulding industry. In Egypt vessels were made during the reign of Thutmose III (1504–1450 BC). Thutmose began a series of Asiatic conquests in

1481 BC and it may be that he brought back workers to set up a glass vessel industry in Egypt. There are similarities in decorative styles, and the extensive use of cobalt in Egyptian glass (as a colouring agent) suggests a Mesopotamian connection, as the nearest source of cobalt was in present-day Iran. Early Egyptian vessel manufacture passed through a long formative stage and it was only in the last decade of the fifteenth century BC that the industry began to grow. In the fourteenth century BC mature industries were in existence in both Egypt and Mesopotamia and glassmaking flourished, but by about 1200 BC peoples from Libya and Asia started to threaten the Egyptian state: internal conditions were also unsettled and the industry there declined. Glassmaking was kept alive, however, by the Syrians, and Syria and Mesopotamia became the two main centres of glass manufacture when revival started during the ninth century BC. Although the products of the two areas seem to have been distinct, it is likely that both industries were influenced by the Phoenicians and were probably staffed by them. From their home along the coast of present-day Lebanon the Phoenicians traded overseas, spreading the products of the glassmakers, amongst other things, throughout the ancient world. Glassmaking centres grew up in Cyprus, Rhodes, and the Italian peninsula: in the fifth century BC either this industry or its products spread to the region around modern Venice and up into present-day Austria.

Glassmaking in Mesopotamia declined at the end of the fourth century BC following the conquests of Alexander the Great, but the Syrian industry flourished, a speciality being plain, moulded monochrome bowls produced in a range of colours. In 332 BC Alexander founded the city of Alexandria, and here the Egyptian glass industry again prospered for the first time since the eleventh century BC. Craftsmen were attracted from a wide region and fine glassware was exported as far as Greece and Italy. Alexandrians may have introduced their new techniques into the Italian peninsula in the first century BC, and the Syrians too established glassworks in northern Italy around the beginning of the Christian era.

Techniques of ancient glassmaking

Prior to the middle of the first century BC glass vessels were made on a shaped core. The core, probably made of mud bound with straw and fixed to a rod, was covered with glass either by dipping it into the molten glass or by wrapping molten threads of glass around it. The surface was then smoothed by continual re-heating and rolling on a flat slab. Surface decoration was added in the form of blobs or trails after which handles and footstands were put on and the core was finally chipped out. Most pre-Roman glass vessels still in existence were made in this way.

The millefiori (thousand flowers) process was used to make beakers, shallow dishes and cosmetic containers. A core was made, of the shape of the inside of the required vessel, and sections of monochrome or polychrome glass rods, loosely held in position by adhesive were laid on the surface of the core. A second mould was placed in position to keep the sections together whilst the glass was fused. The moulds were then removed and the surfaces of the vessel ground smooth to produce a fine mosaic effect. The Alexandrian workshops were famous for these mosaic products, which may have been introduced by craftsmen from western Asia where the process was developed.

Glass was cut or ground, by harder materials such as quartz, into pots and ornaments. It was also pressed into bowls using open moulds of fired clay, and shaped between two parts of a closed mould by fusing powdered glass *in situ*. More complex was the lost wax process where a wax replica of the required glass object was coated with clay which, having some strength whilst still unfired, held the shape when subsequently warmed to melt out the wax: the clay was then fired to form a mould into which the glass could be poured.

All these techniques were in use to provide a wide variety of useful and decorative objects, but the picture changed dramatically with the first great revolution in glassmaking, the invention of glassblowing.

The discovery of glassblowing and the growth of glassmaking in Europe

It used to be thought that the invention of glassblowing occurred in the decade immediately preceding the Christian era until a find in 1970, from excavations in the Jewish quarter of the old city of Jerusalem, of a sealed deposit of cast and blown glassware. This included large amounts of waste products from a workshop producing blown vessels, dating from 40–50 BC. These finds indicate that glassblowing originated in the eastern Mediterranean, if not along the

Phoenicio-Syrian coast, several decades earlier than previously suspected.

The glassmakers used a hollow rod, on the end of which they collected a sufficient amount of glass to make the required object. They then blew down the rod to form a glass bubble. Glass hardens gradually as it cools, and this makes possible free, or off-hand blowing to form a wide variety of shapes. The bubble of glass could also be blown inside one-, two- or three-part moulds to yield many complex forms. Compared with the limited shapes of core-formed vessels objects were produced which could be used for a much wider range of applications, and, moreover, could be made more cheaply and quickly. The first age of mass production of glassware had arrived.

Conditions were suitable for the exploitation of this new technique. The Roman hegemony over the Mediterranean area was being consolidated, leading to relative stability and prosperity. The main centres of glassmaking in the Hellenistic world fell under Roman control during the first century BC and there were also profitable markets in newly conquered provinces of northern and western Europe.

The speed of adoption of the new technique was very rapid: for example the initial period of Roman occupation at Cosa in Etruria, 273–20 BC, yielded only 30 coremoulded and cast vessels from excavations, whilst hundreds of fragmentary vessels, mostly blown, were found for the period 20 BC to 30 AD. Archaeological evidence from numerous Italian sites indicates that blown vessels first appeared in Italy during the last quarter of the first century BC. Under Augustus, Rome was considered a major glassmaking centre. Then glass became utilitarian and inexpensive, and could be purchased by a wide section of the population. Indeed, glass was so common that glassware upon a Roman table was a sign of lack of affluence, the rich using gold and silver plate. This expansion brought its problems: from 200 AD the Roman city fathers forced glassmakers to concentrate in the suburbs away from the city because the pollution caused by the numerous furnaces had become so troublesome.

The Roman glassmakers did not make only vessel glass: window glass first came into widespread use during this period. Pompeii, a city famous for its luxurious style of living, boasted windows glazed with large sheets of glass. The bath house windows, for example, were of thick glass measuring about 40 by 30 inches. Much Roman window glass was of a greenish-blue colour, small pieces being fitted into a more or less richly ornamented wooden frame divided into many sections. It was probably cast as blocks, the hot glass being poured or pressed into flat open clay moulds or even poured out upon flat stones.

Glassmaking was not restricted to Italy and the eastern Mediterranean area: the northern regions of the Roman Empire also had their glassmaking areas. Indeed, it was at the important centre of Trier that the late-Latin term "glesum" may have originated, probably derived from a Germanic word for a transparent, lustrous substance: hence our modern word, glass. Factories such as those at Trier and Cologne were famous and prospered for several centuries. There was also continual movement of workers within the Empire, so that it is often impossible to tell where a particular piece was made.

Glassmaking after the decline of the Roman Empire

The existence of widely dispersed centres of glassmaking helped the glass industry to survive when the Roman Empire started to decline. Glassmaking continued in the valleys of the Rhine and the Rhône, although many workers fled to Italy, particularly to the Po valley and to Altare, near Genoa (whence they later spread out all over Europe). Although the industry as a whole survived, the art of making vessel glass went into a decline. The new Teutonic patrons probably demanded simpler, plainer shapes and knowledge of many complex Roman techniques was lost. It is interesting to note, by way of contrast, that in the East the industry continued to flourish after the Islamic conquests, and the art of painting, enamelling and gilding of glass was developed. This fine art continued until the Mongol conquests drove large numbers of glassmakers from Damascus (sacked by Tamerlane in 1402) and Aleppo. Many came to the West where they were to have a great influence on glass design.

The ingredients used by the early western glassmakers were the same as those employed by the Romans: sand containing shell remains probably provided silica and lime (calcium oxide) whilst the soda (sodium carbonate) which was used as a flux to lower the viscosity of the resultant glass would be present in the ash of sea plants, obtained from the Mediterranean coasts. Towards the end of the tenth century AD there was a fairly rapid switch in the north-western area of the former Roman Empire to using potash (potassium carbonate) as a flux, obtained from bracken and other woodland sources such as beech wood. The glassmakers could easily obtain these materials on their travels, as the woodlands were far more extensive then than they are today and glassmaking centres were established throughout such wooded regions as Germany and Bohemia. After the tenth century this potash glass became characteristic of central Europe whilst soda glass continued to be made in the coastal regions.

The patronage of the Church

The production of window glass in the West does not seem to have suffered in the same way as that of vessel glass after the fall of the Roman Empire, and this was in large part due to the support of the Church. The Church, with its near monopoly of production and collection of manuscripts, acted as a recorder and preserver of the methods of glass manufacture. It was also a rich patron, needing glass for the windows of its buildings in great quantities. Bede tells us that in 675 AD the abbot Benedict Biscop sent for glassmakers from Gaul to make windows and vessels for his new monastery at Monkwearmouth, and in 758 AD Cuthbert, abbot of Monkwearmouth and Jarrow requested the services of a glassmaker from Mainz. This documentary evidence is supported by the finding of fragments of clear and coloured window glass at both these sites. Glastonbury, another Benedictine foundation, has also yielded evidence of actual glass manufacture, dating from the ninth or tenth century AD, and so it is reasonable to assume that the industry was even more widespread on the Continent.

By the end of the tenth century AD conditions in Europe were becoming more settled and church building began to flourish. However, the techniques of building were such that window areas were relatively small. In the early twelfth century AD, the introduction of the Gothic style with its pointed arch and, later, flying buttresses, took much of the stress off the walls, enabling the builder to construct larger windows, which they filled with panes of glass in vivid colours. These windows not only added to the beauty of the churches, glowing like jewels within the dark interiors: they also fulfilled theological and didactic purposes, reflecting the idea of God as the source of perfect light and explaining the Bible stories in simple and dramatic pictures to people who could not read.

The beauty of the old glass lies partly in its design and partly in the material itself. It is in a large part due to its imperfections: bubbles and striations cause variations in refractive index which result in a richness not usually present in more perfect modern glass. Also, the colouring agents used were less pure, and this can give a less harsh appearance to the glass.

I have included some of the old glassmaking texts in the bibliography at the end of this chapter. They throw a great deal of light on techniques which were in use for hundreds, if not thousands, of years. The people who wrote them lacked our modern scientific knowledge, and so could only describe processes in empirical terms: often they could not understand the phenomena that they observed and so their descriptions can be difficult to follow. Nevertheless it is possible to discover a great deal from them.

The process of constructing a window must have been a lengthy one: the workers had little control over their raw materials and it would often have been difficult to produce a piece of glass of the right shade for a particular purpose. In modern terms only a few elements were used as colouring agents, but they were obtained from many sources and their preparation was complex. Blue, one of the most important colours in medieval glass, was obtained from "zaffre", an Arabic word for cobalt oxide: the material containing this oxide was very expensive as it had to be imported from the Levant and so was known as Damascus pigment. Later, supplies were obtained on a large scale from Saxony. The cobalt was extracted by roasting the ore so as to remove sulphur, arsenic and other volatile matter. Another compound used to give blue was the residue left behind when bismuth was separated from its ore. This residue must have contained cobalt with traces of nickel, and probably of iron and copper, and the blue produced would differ from that due to pure cobalt. Combinations of zaffre and copper compounds (e.g. calcined, or strongly-heated, brass) gave a sea-green.

Copper and iron compounds were used to give greens and reds. Copper was generally added as ferretto, or burnt copper, made by a recipe dating from classical times. The copper was heated with sulphur to give a black mass of copper sulphide which was then roasted until it was converted into the red ferretto. Alternatively the copper was heated with blue vitriol, or copperas (copper sulphate), the resulting material probably containing a large proportion of basic copper sulphate. Copper could also be added by the inclusion of calcined brass, or brassmakers' scales, the mixture of oxides of copper and zinc formed when brass was heated; "crocus martis", ferric oxide, made by the same process as ferretto, but using iron instead of copper, was widely employed for yellows and browns. As with many medieval materials, the words "crocus martis" covered many reddish compounds, e.g. ferric acetate, nitrate or chloride made by treating iron filings with vinegar, aqua fortis (nitric acid) or aqua regia (a mixture of nitric and hydrochloric acid). Opaque white glass was produced by tin oxide, and purple by manganese.

Deep red, the colour most frequently used after blue, could be produced by iron mixed with a little calcined brass but the finest reds or "rubies" were made with copper. Copper ruby glasses could be produced by fusing glass containing copper with a small amount of tartar (potassium hydrogen tartrate) which reduced the copper to the cuprous state; the reducing atmosphere of the old furnaces also had this effect. On reheating the glass a colloidal dispersion of copper and probably of cuprous oxide was produced, giving a fine red colour. This colour was so intense that window glass had to be "flashed": a bulb of clear glass was dipped into a pot of copper ruby glass to give a thin transparent layer on top of the clear glass.

Details of a scene on a stained glass window, such as the folds of drapery, were applied by means of painting, and then firing, a black enamel pigment derived from iron. From the thirteenth century a second pigment in the form of a "stain" of silver chloride or sulphide was used. When the stain was applied to the clear glass and fired colours varying from yellow to orange were produced. The stain was later applied to blue glass to give a green colour, making possible the depiction of blue sky and green fields on the same piece of glass.

Glass is one of the most fragile materials used in buildings and so it is not surprising that relatively little has survived until the present day. Even so, we might expect to find more medieval glass than we do, judging from the evidence of churches and cathedrals all over Europe, and the many pictures which show glass being used for a wide variety of everyday purposes. There seem to be two reasons for this. Firstly, until quite recently excavators were mainly interested in the architectural aspects of buildings and they ignored fragments of glass that were unearthed. Even if they took note of these, they did not record their position and stratification, making dating very difficult. Secondly, the potash glass of the medieval glassmakers is much more prone to weathering and disintegration in western European soils than is the soda glass of the Roman period (see Chapter 5) and so most of it must have been destroyed long ago.

Methods for making window glass

The glass which was used for complex and beautiful window designs was made in two ways, as crown glass or as broad glass. The origins of crown glass date back to Roman times, although the technique was developed by the glassmakers of Normandy and it was in general production until the end of the eighteenth century. The process is shown in Fig. 5.

The workman gathered molten glass on the end of the blow-pipe, then blew down the pipe to expand the glass to form a globe. As this was a lengthy process the glass was reheated from time to time at the furnace mouth to prevent setting. A solid iron rod, known as a "pontil" rod, was then attached to the globe on the side opposite to the blowpipe, which was then cracked off leaving a small opening. The glassmaker took the globe on the rod to the furnace to reheat it, at the same time continuously rotating the globe to prevent it sagging. As the globe was heated the glass became less viscous and at a certain point the centrifugal force caused the globe to "flash", or open out to form a large flat disc, centred on the pontil rod. The best pieces of glass from this disc, cut from its edges, were quite small because of limitations on the size of the disc, but they could be made very thin (a desirable characteristic, for the glass was often heavily coloured by unavoidable impurities in the raw materials) and the glass surface, untouched during manufacture, had a bright fire-polish. The thicker "bull's-eye" at the centre of the disc was not wasted: sold more cheaply they can often be found in the windows of domestic buildings. Modern imitations are often made by allowing the softened glass to sag into a mould: they can be distinguished by the fact that the glass does not vary in thickness, the older glass having a lenticular shape in cross section.

Broad glass making also changed little over the years. The process was known at least as early as the twelfth century AD, when it was described by the German monk Theophilus in his book *Schedula Diversarum Artium* and was still in use during the eighteenth century. Briefly, an elongated bulb of blown glass was slit along one side and flattened out to form a sheet. The process is shown in Fig. 6.

Broad glass could be made into larger pieces than crown glass but as the surface of the sheet came into contact with the tools during manufacture it lacked the fire-polish of crown glasss.

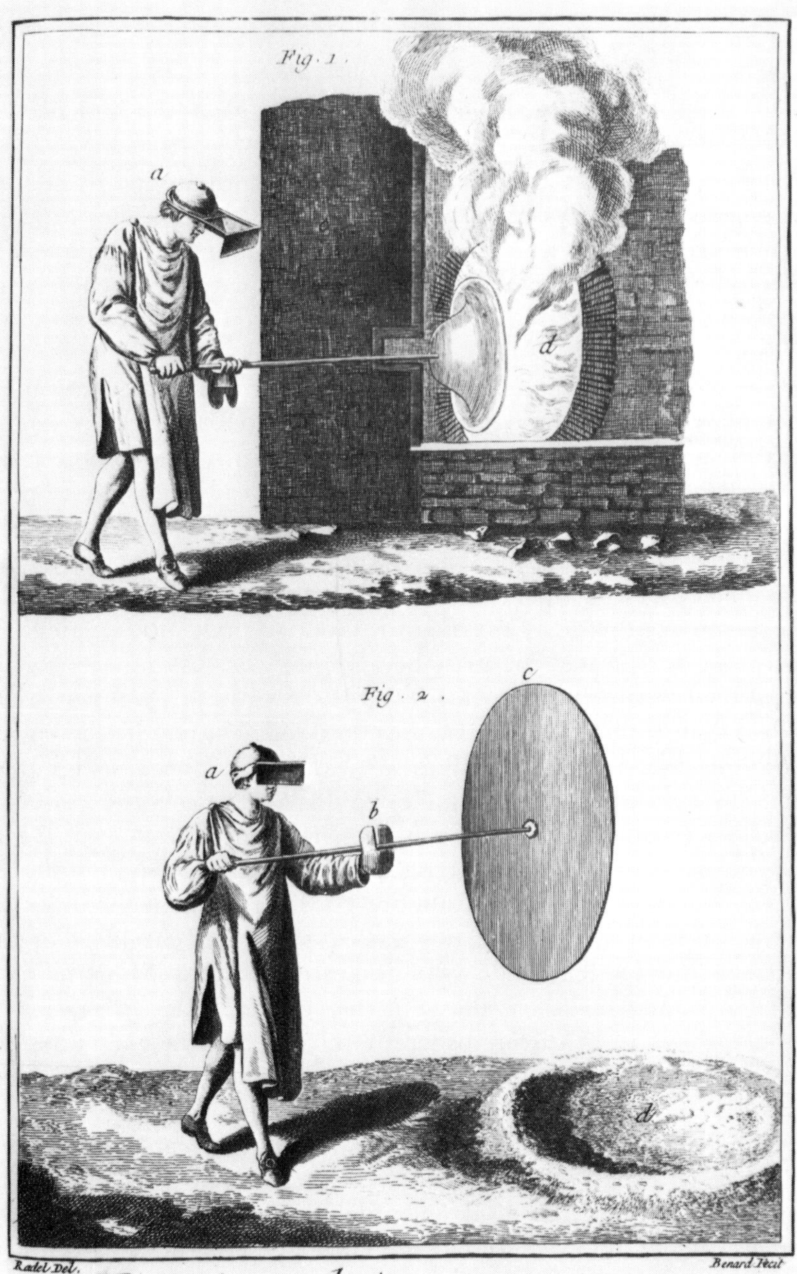

Fig. 5. Making crown glass: an eighteenth-century illustration.

Broad glass was the speciality of the Lorraine glassmakers. They came originally from Bohemia, but in the early fifteenth century they started to travel, settling eventually in the wooded regions of the Vosges, where they had an ample fuel supply. A fifteenth century Bohemian manuscript shows what their forest glass houses must have looked like: the furnace is rectangular, typical of the furnaces used in northern Europe at that time (Fig. 7).

The other type of furnace, originating in Mediterranean countries but later spreading through Europe, was of a "beehive" shape. Because of its origins it was known as a "southern" furnace, as opposed to the rectangular "northern" furnace. A version from an eighteenth century drawing is shown in Fig. 8.

Many variations on these structures are known, but they all had a space for the fire below the melting compartment, where the glass was melted in pots. The fire often served to heat another compartment reserved for annealing the formed glass objects, a process which was necessary to prevent the finished articles from shattering (see Chapter 1).

The furnaces also had compartments for a process known as "fritting". Much ancient glass is opalescent or opaque, and contains many bubbles. This is often due to the fact that the furnace temperatures were low in comparison with modern furnaces: from a study of the composition of the furnace building materials it has been estimated that the temperature did not normally reach much more than 1100°C, and extensive improvements were not made until the nineteenth century. The fritting process was an attempt to improve the quality of the glass by pretreating the raw materials before melting them. They were placed in the fritting compartment, at a lower temperature than the melting section, and raked over from time to time to expose fresh surfaces to the heat. This eliminated some of the gaseous products which otherwise would have remained in the glass, and burnt off various impurities. As it was a lengthy process, many of the necessary glassmaking reactions had time to occur. The "frit" was then water-quenched to break it up and to wash away impurities, ground, well mixed, and finally melted to give a glass of reasonable quality.

The Venetian glass industry

We have concentrated so far on the production of glass in the more northerly regions of Europe, where France, with its vast beech forests,

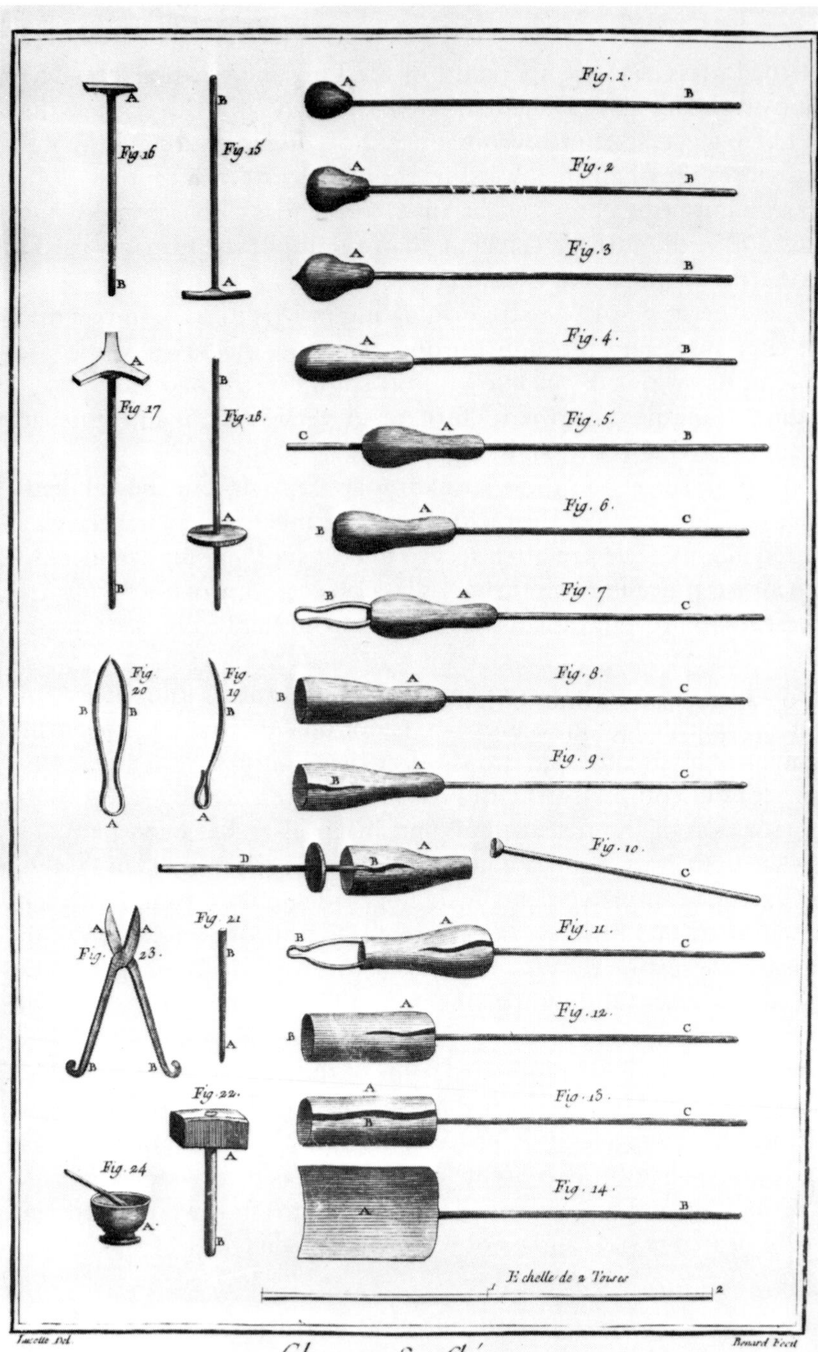

Jacotte Del. Benard Fecit

Glaces Soufflées,

Opérations progressives et Outils.

was a major centre of the medieval window glass industry. However, towards the end of the fifteenth century the centre of interest shifted to Venice, which then became a dominant influence in European glassmaking for the next 200 years.

Glass was made in Venice from the tenth century, but in 1291 the glassmakers had to move to the island of Murano, because of the fire risk in Venice itself. This isolation and concentration favoured the development of a powerful guild of glassmakers, which had both good and bad effects: the glassmakers enjoyed many privileges and could experiment with new designs and compositions, but they were not allowed to take their secrets from the island. Those who did so were often hunted down and killed, but in spite of this persecution Venetian glassmakers helped to introduce the Italian style throughout Europe.

The rise of Venice as a glassmaking centre was due to two main factors. Until the fall of Constantinople the Venetians were in a favourable position for trade with both western Europe and the Byzantine Empire. After 1453 their eastern markets dwindled so, relying on trade for their prosperity, they had to increase their exports to the West. At the same time they managed to develop a glass as clear as rock crystal which could be blown very thin, known as "cristallo". They also discovered how to produce very good glass in a variety of colours and to decorate with gilding and enamels, techniques probably learnt from Eastern refugees. Their beautiful designs, both ornate and delicate, were the basis for a style, the *façon de Venise*, which spread rapidly, along with their own glass exports, to other European countries. Although it was adapted to the taste of different regions, the Venetian influence remained predominant until the end of the seventeenth century.

Glassmaking in England

Although extensive finds of both Roman and Anglo-Saxon glass have been made in England, it is uncertain how much of it was produced in this country. However, glass was made here in medieval

---

Fig. 6. Making broad glass:an eighteenth-century illustration. A bulb of glass is blown into an elongated shape, opened out at each end in turn, and finally slit along its length and flattened out to form a sheet.

Fig. 7. A Bohemian forest glass-house of the fifteenth century. In the background a man digs sand from the hillside and fuel is carried in a basket. In front are the glassblowers gathering glass and blowing a vessel, whilst a boy tends the furnace and the worker on the left removes the vessels for annealing. Reproduced by courtesy of the Trustees of the British Museum.

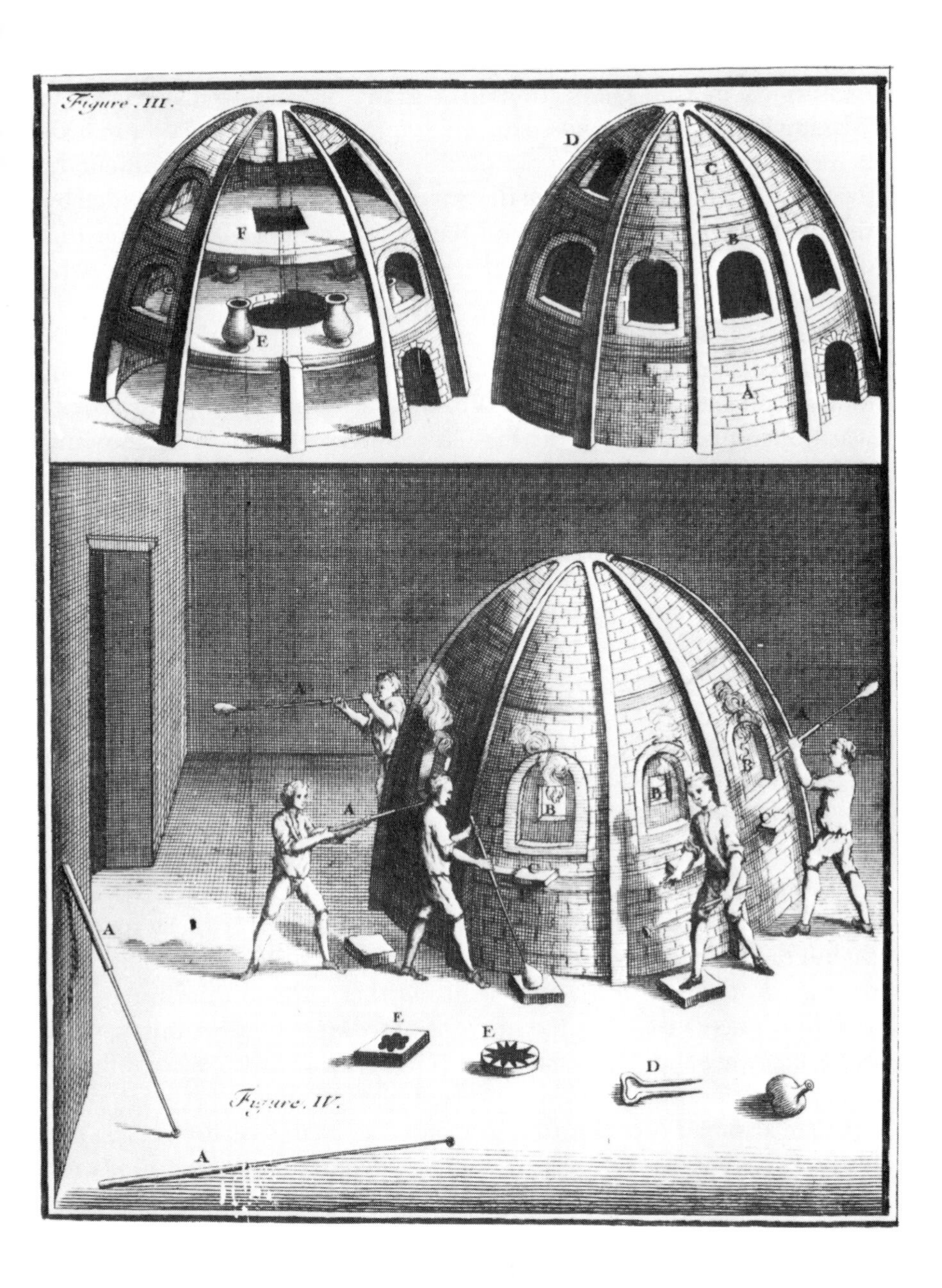

Fig. 8. A "southern" type glass furnace. It has three compartments, the lower one for the fire, the middle one for the pots containing the molten glass (worked through holes in the walls), and the upper one for annealing the finished articles.

times, the chief centre being Chiddingfold in Surrey. The area is first mentioned in this connection in a grant of land to Laurence the Glassmaker, dated around 1226, and fragments of fused glass and broken crucibles have been found on this site. Conclusive documentary evidence of a glass industry in the Surrey–Sussex Weald is provided by the Exchequer rolls in 1351, and the picture can be filled in after this date from accounts, parish registers, deeds and parliamentary petitions.

The medieval English glass industry was probably started by glassmakers from France, although other glassworkers came from long-established families in the Weald. The industry was rather backward in comparison with its French counterpart, however, showing little expansion until the third quarter of the sixteenth century. The quality of its wares seems never to have matched that of Normandy glass, which was imported on a large scale. Nevertheless the industry survived in a quiet way: wood for fuel and other raw materials were plentiful and the manufacture, requiring little in the way of capital equipment, could often be fitted into the farming year as a source of extra income.

England was thus an area ripe for development, with a virtually untapped market. Taking advantage of this situation, a merchant from Arras in Flanders called Jean Carré obtained a monopoly licence in 1567 from the Government of Queen Elizabeth I for making window glass in the Weald. He has been credited with the introduction at this time of skilled glassmakers from Lorraine. Although recent evidence throws doubt upon this connection, the Lorrainers were very successful but they soon came into conflict with the native population, largely owing to competition with iron workers for diminishing fuel supplies. This caused them to move to other areas of the country, such as Stourbridge and Newcastle-upon-Tyne, where they were influential in developing many new centres of glassmaking.

Carré also had the intention of making crystal glass for vessels but his workers were not experienced in this manufacture and by 1571 records show that he was associated in London with a group of Italian glassmakers, presumably skilled in the art of producing crystal glass. However, it is doubtful whether he was successful and when he died his place was taken by a Venetian, Jacob Verzelini. In spite of many reverses, such as the malicious burning of his glasshouse, he obtained in 1575 a monopoly to make Venetian style glasses and a 21 year prohibition on imports. This prevented the market from being flooded by cheap foreign glass and together with his technical competance and business ability formed the basis for the success of the British glass industry during the next two centuries.

Coal-fired furnaces

As the glass industry grew it became increasingly difficult for the glassmarkers to obtain adequate supplies of wood for firing their furnaces. They used large quantities of fuel and were in competition with other manufacturers, notably iron workers, for the diminishing supplies. Also wood was needed for the construction of ships for the Navy, and the Government was very concerned that the stocks of timber were diminishing. These factors led to experiments in the heating of furnaces by coal in the early seventeenth century and in 1610 an agreement was made between the Crown and a certain William Slingesby, by which the latter was given the sole right to use, in various industries, coal in place of wood. In 1611 a Crown monopoly was granted to a small group of gentlemen for making glass using coal as fuel, and in 1615 an official decree was issued prohibiting the use of wood for the firing of glass furnaces, and insisting on the use of coal.

The growth of the industry at this period was largely due to the efforts of one man, Robert Mansell. He was a member of the 1611 monopoly group, but by 1618 he had gained complete control, and for 38 years he dominated the industry. Glassmaking now developed in coal-producing areas as Mansell established groups of glassmakers close to the coal fields in southern Scotland, around Tyneside, and at Stourbridge. Stourbridge also had the advantage of an excellent supply of fire-clay, used in the production of pots for glass melting.

Glass was not the only commodity on which monopoly rights were granted under James I and Charles I. In efforts to raise money they established monopolies on almost every article of domestic consumption, and this usually increased prices out of all proportion to the profit gained by the Crown. Thus although Mansell's enterprises flourished there was no scope for healthy competition and new developments by others until his death in 1656.

Lead crystal glass

The opportunity for fresh advance in the English glass industry coincided with the decline in popularity of the elaborate Venetian styles which had dominated European taste for so long. Starting in the second half of the seventeenth century various types of glass were developed in England, culminating in heavy lead crystal glass which was admirably suited to the new simpler styles.

Although lead had been used since ancient times in glass, crystal glass containing lead was developed by George Ravenscroft, who started his experiments in London in 1673. From the start his glass had a better appearance, probably owing to his care in selection of raw materials and in 1674, having obtained a patent, he was set up in a glasshouse at Henley-on-Thames, Oxfordshire, by the Worshipful Company of Glass Sellers of the City of London. Here he could continue his experiments while the Company undertook to find a market for his whole production, provided that the vessels were made in specified shapes and sizes.

Ravenscroft's original glass may only have been a reproduction of the Venetian "cristallo" but before 1676 he was probably adding lead oxide to the batch materials, and he was certainly using nitre (potassium nitrate) and possibly potash as fluxing agents, rather than soda. (The use of a potassium rather than a sodium compound as the alkali in lead glass is preferable for the production of high quality blown glassware.) At first he had difficulties due to "crizelling": in the absence of lead oxide the large amounts of potash or nitre required to flux the silica gave a glass readily attacked by water resulting in a network of fine surface cracks. Further work on the proportions of the constituents seems to have been successful and in June 1676 the Glass Sellers Company announced that vessels made from the material would be satisfactory. By 1685 the production of the new glass was well established and the members of the Glass Sellers Company were selling a wide range of fine quality glasses. Very few certain examples of Ravenscroft's glasses are still in existence, but tests on pieces reasonably attributed to him give a lead content of about 15%.

Following the success of lead crystal glass, the British glass industry achieved a leading position in the eighteenth century which it was to hold for 100 years. The beauty of the drinking glasses of this period has rarely been surpassed. They show a great variety of shape, size, decoration and stem type: air twist, opaque white or coloured twist (Fig. 9). Although Britain was a leading manufacturer every country in Europe was making more glass: glasses even passed between countries before completion, such as those goblets sent from England to Holland for engraving prior to re-importation and sale.

Fig. 9. Eighteenth-century English wine glasses of lead crystal. The clarity and brilliance of the glass is enhanced by the elegant designs, delicate twisted stems and fine engraving. Reproduced by courtesy of the Department of Ceramics, Glasses and Polymers, University of Sheffield.

German and Bohemian glass

Although lead crystal glass gained for Britain a leading place in European glassmaking important developments elsewhere introduced new styles which competed with the prevalent Venetian fashions. Elaborate carving on rock crystal had long been a speciality of the Bohemians and towards the end of the seventeenth century they developed a heavy, clear potash-lime glass similar in appearance to natural rock crystal which could be blown into substantial vessels and then decorated in a similar manner. The glass, the skill of the cutters and the artistic climate of the time combined to produce fine pieces which were sold by well-organized marketing groups throughout Europe.

Coloured glasses were also developed to a high degree by the Germans. A notable experimenter of the time was Johann Kunckel, the director of the glasshouse of the Elector of Bradenburg in the 1670s, who was the first to develop and describe a reliable formula for the preparation of rich, ruby-red glass, using gold chloride as the colouring agent. He also produced an opaque white "porcelain glass" made by adding burnt bone or horn, which contained phosphates, as the opacifier. This became very popular during the eighteenth century when there was a vogue for objects made in imitation of true porcelain.

Glassmaking in North America

The eighteenth century was also an age of expansion for the American glass industry when a wide range of goods was produced. Glassmaking was first established there in 1608 within a year of the founding of the earliest English settlement at Jamestown, Virginia, but it did not prosper and during the seventeenth century various other enterprises met with little success. The first to survive for a longer period was the one founded by a German, Caspar Wistar, in 1739 in Salem County, New Jersey, which he and his son operated until the outbreak of the Revolution. They produced sturdy, free-blown glass which reflected the independent, puritanical outlook of the colonists in an area where Quakers formed the majority of immigrants.

No domestic competitor of any importance appeared until 1763, when Henry William Stiegel, a colourful character popularly known as "Baron" Stiegel, working with a team of German glassblowers, started to produce a range of glassware for the domestic market. However, Stiegel had great ambitions to make fine crystal on the European pattern. He built up a team of more than 130 workmen and brought to America engravers, painters and glassblowers skilled in Venetian techniques, also glassmakers from Bristol. Although some beautiful glass was made the venture was not successful for long and Stiegel went bankrupt in 1774. The only other enterprise to make large quantities of fine glassware for domestic use at this time was the New Bremen factory in Maryland, run by Johann Friedrich Amelung. Other glassworks were probably on a much smaller scale, making mostly bottles and window glass, but the industry expanded sufficiently during the last two decades of the century to establish itself as a successful American manufacturing process. Skilled workers were brought over from Europe and the local men learned their skills, sources of raw materials were located and tested, and thriving works were established along the eastern seaboard and also west of the Alleghenies. These successes were to prepare the way for the development of the industry during the nineteenth century when North America became a leader in an age of mass production.

Expansion and automation

By the end of the eighteenth century expanding economies in Europe and America were increasing the demand for glass. The establishment of large, well-equipped and adequately financed factories was accompanied by the development of well-organized sales and distribution. At the same time the Industrial Revolution was taking place and the science of chemistry was beginning to develop. This in turn lead to the growth of a chemical industry, of which a considerable part was concerned with production of alkali for the glass and soap manufacturers. The alkali could be turned out in large quantities, with reasonably constant composition and a continuing guarantee of supply, all of which were necessary for the production of glass on a large scale to satisfy an expanding market. Improvements also occurred in other batch materials: for example, sands were carefully selected for their purity, including freedom from iron, very small amounts of which give glass a strong green colour. These trends have continued to the present day, when pure materials with accurately known compositions are essential to the modern, scientific production of glass.

The increased availability of better raw materials, improved organization of glassworks, and more successful exploitation of markets would have been of little use without the development of machines for mass production. It was in North America, where workers were scarce and wages were much higher than in Europe that means were urgently sought to increase productivity. By the middle of the nineteenth century acceptable, cheap, pressed glassware was produced there on a massive scale for general consumption as a substitute for the expensive cut lead crystal glass.

In contrast, as late as 1890, the rest of the American glass industry was still basically a craft industry. Their shops had less efficient furnaces and used older processes than, for example, those in England, where great strides had been made in the large-scale production of window glass (which, however, still required a good deal of handcraft). Many reasons have been suggested for this: they include shortage of labour with sufficient skill and technical knowledge, reluctance to adopt new methods, high import tariffs which protected the inefficient from competition, and processes which did not compel production on a large scale. Nevertheless, by the 1880s, glassmaking was approaching a technological revolution. The basic features of precision machinery had been developed. Tank furnaces, in which glass could be melted continuously, on a large scale (as opposed to pot melting which was intermittent and small-scale) were available. Natural gas was now produced on a large scale and temperature control, vital for the automatic forming of glass, had been made easier by the substitution of gaseous for solid fuel.

These great opportunities were recognized and exploited by a relatively small number of men whose activities centred on Toledo, Ohio. Edward Drummond Libbey, a man of outstanding enterpreneurial skill, and Michael Joseph Owens, an inventor of genius, combined to introduce processes that eventually rendered obsolete most of the industry's technology throughout the world. Owens and his fellow workers developed bottle-making machines, machines for the production of lamp bulbs, tube-drawing machines and window glass machines. By 1920 the hand workers in most branches of the industry had been displaced and other manufacturers had been driven to develop new processes which further accelerated the trend towards mechanization. In less than 50 years the foundations of the glass industry as we know it today had been established.

Bibliography

This chapter is a survey of glass history and is therefore necessarily superficial in its coverage of seven thousand years. There are many books and articles on this history, the following being a personal, and limited, selection. They are mostly of a general nature: for specialized texts on a particular country or period you should consult one of the indexes mentioned in Chapter 8.

Vose, R. H. (1980). *Glass.* (Collins)

A survey of glassmaking history and technology from its beginnings to the twentieth century giving particular emphasis to the archaeological evidence for the development of British glassmaking, and offering a guide for the amateur archaeologist on the techniques of glasshouse excavation. This text will form a useful companion volume to the present work, as its contents are in many ways complementary.

Barrington Haynes, E. (1959). *Glass through the ages.*(Penguin)

Although published in 1959, this book is still one of the best introductions to the history of glass. In the second part of the book English drinking glasses of the eighteenth century are classified and described in detail, and there are many line drawings and photographs.

Kämpfer, F., and Beyer, K. G. (1966). *Glass: a world history.* (Studio Vista)

This book tells the story of glassmaking, in pictures, from its origins to the present day. There are 243 fine plates, 40 of them in colour, with detailed captions, showing outstanding examples of glassmaking from every period and every part of the world. There is a helpful glossary and index of technical terms.

Weiss, G. (1971). *The book of glass.* (Barrie and Jenkins)

A history of decorative glass in the Western world from earliest times to the present. An excellent survey containing over 500 illustrations, many in colour. Notable features are the line drawings showing the development of styles and the many clear chronological charts.

Polak, A. (1975). *Glass: its makers and its public.* (Weidenfeld and Nicolson)

The first full social history of glass and glassmaking, covering the period from medieval times to the industrial age. A very good overview of a complex subject.

Douglas, R. W., and Frank, S. (1972). *A history of glassmaking.* (Foulis)

A general survey of glassmaking which concentrates upon the neglected area of technological history.

Charleston, R. J., *et al.* (eds) (1969). *Studies in glass history and design: papers read to Committee B sessions of the VIIIth International Congress on Glass held in London 1st–6th July, 1968.*

The International Congress on Glass, held every three years, has since 1974 been mainly concerned with the scientific and technological aspects of glass. Prior to this date the published proceedings, which include review articles and a section of papers on glass history and design, provide valuable material for archaeologists. This book, published as a separate volume of proceedings, contains many interesting papers.

Harden, D. B., *et al.* (comps.) (1968). *Masterpieces of glass.* (Trustees of the British Museum)

This is the catalogue of a unique exhibition which consisted of outstanding pieces of glass, entirely from the collections of the British Museum, and ranging in date from Egyptian glass of the fifteenth century BC to European productions of the eighteenth century AD. Each section is preceded by an introduction, written by an expert in the glass of the period, and the 269 objects on display are all illustrated. There is an extensive bibliography.

Oppenheim, A. L., *et al.* (1970). *Glass and glassmaking in ancient Mesopotamia.* (The Corning Museum of Glass)

A scholarly study, by four leading experts on glass, of the cuneiform texts of instructions to glassmakers, their chemical interpretation and their relationship to other glassmaking areas in the ancient world.

Kenyon, G. H. (1967). *The glass industry of the Weald.* (Leicester University Press)

An authoritative and detailed account, based on personal experience, of the glasshouse sites of the Wealden glass industry which flourished in the Surrey and Sussex Wealden area from the thirteenth to the seventeenth century. I include this book both because of its archaeological aspects and because it describes glassmaking as it must have been carried out in many places in Europe in the pre-industrial age.

Godfrey, E. S. (1975). *The development of English glassmaking 1560–1640.* (Clarendon Press)

A detailed economic and political study of English glassmaking at a crucial point in its history, when it was being transformed from a medieval craft into something resembling a modern industry.

Scoville, W. C. (1948). *Revolution in glassmaking: entrepreneurship and technological change in the American industry 1880–1920.* (Harvard University Press)

The title of the book is slightly misleading, because it provides an account of the American glass industry from its beginnings in the days of the first European settlers. This is a detailed and meticulous survey which considers the economic, technological and social factors involved in the development of the American glass industry. It describes, using the outstanding example of American innovation, the change from small craft enterprises to modern methods of mass production.

Halliday, S., *et al.* (1976). *Stained glass.* (Mitchell Beazley)
It would be a mistake to pass this over as a coffee table book, although the main feature is the fine colour illustrations, 500 photographs providing a guide to stained glass windows from all over the world. The text places stained glass in its historical, social and artistic contexts. Manufacture and aspects of restoration are also considered. A gazetteer lists some important European and American stained glass, from the Middle Ages to the present day.

For those who want to study stained glass in more detail, the *Corpus Vitrearum Medii Aevi* has since 1956 been overseeing the production of various series of country volumes which aim to record, in over 70 volumes, stained glass still extant in the particular country. The glass is fully documented, illustrated and indexed.

If you need to keep up to date with the latest work on glass history you should follow the recommendations for literature searching given in Chapter 8 but mention may be made here of some special serials devoted to glass history.

Journal of Glass Studies (Corning Museum of Glass, 1959–)
This is a very important source of authoritative original papers on all aspects of glass history.

Engle, A. (ed.). *Readings in Glass History*. (Phoenix Publications, 1973–)
The journal includes papers on glass history which aim to place glassmaking in its wider historical and social contexts. Volume 9, 1977, is an index to volumes 1–8, and volume 10, 1978 is an illustrated companion to volumes 1–8 covering glass history to the end of the Roman-Byzantine period.

The Glass Circle. (Oriel Press, 1972–)
This journal appears at irregular intervals and contains papers contributed by the Circle of Glass Collectors. The subjects are wide-ranging and often provide information on topics which would not otherwise come to the attention of students of glass.

Finally, a selection of old glassmaking texts. These add an extra dimension to the study of glass and repay careful study. They do not replace individual documents of the period but they do summarize the "state of the art".

Dodwell, C. R. (ed.). (1961). *Theophilus:* De Diversis Artibus (*The various arts*). (Nelson)
This is the translation of a medieval treatise written by a Benedictine monk during the first half of the twelfth century. The author was a prac-

tising craftsman rather than a theologian, and an artist of some standing and experience. Although his primary interest was probably metal-work, Book II of the treatise is devoted to glass. Theophilus describes processes in some detail. Subjects covered include furnace construction, preparation of raw materials and pots for melting the glass, production of coloured, vessel, and window glass, and the construction of stained glass windows. The editor provides a detailed introduction.

Smith, C. S., and Gnudi, M. T. (trans.) (1943). *The* Pirotechnia *of Vannoccio Biringuccio*. (M.I.T. Press)

Originally printed in Venice in 1540, this is a comprehensive treatise on "the arts of smelting or casting metals and all related subjects", including glass. Biringuccio describes in detail the construction of the melting furnace and the preparation and melting of the glass. The work is illustrated with line drawings, including one of a glass furnace.

Hoover, H. C., and Hoover, L. H. (trans.) (1912). *Georgius Agricola:* De Re Metallica. (Dover)

First published in 1556, this is the first book on mining to be based on field research. It remained the only authoritative work in this area for nearly 200 years. In his remarks on glass Agricola copies Biringuccio but adds some important variants on the practice. The very clear illustrations of glass furnaces are especially helpful.

Neri, A. (1612). *L'Arte Vetraria*

Antonio Neri was a practising glassmaker who published the first systematic account of the preparation and treatment of raw materials for glassmaking, together with directions for melting a wide variety of glasses. During the next 200 years this influential work ran through many editions, being annotated and translated into several languages. Probably the version of most interest to English-speaking readers is the 1662 version, translated by a distinguished founder-member of the Royal Society, Christopher Merrett. Merrett added to the book his own "Observations" of length approximately equal to the text which he translated. They give valuable information on glassmaking in England at that time. A facsimile copy of this edition has been produced by University Microfilms Ltd.

Because this work is so important it may be helpful to mention here an article which helps to clarify the procedures described in the light of modern scientific knowledge. This is the paper by W. E. S. Turner (1962); a notable British seventeenth-century contribution to the literature of glassmaking (*Glass Technology* **3**, 201–213).

3

Scientific Analysis of Glass Remains

The study of glass remains involves the use of techniques which are in widespread use in other areas of scientific research. The aim of this chapter is to describe the most important techniques, and to show, by means of selected case studies, how modern techniques can provide essential information on glass for the archaeologist.

Information from simple tests

Although most of this chapter will be devoted to studies involving the use of advanced research facilities, a great deal of information can be gleaned from simple observation and tests. I am indebted to Professor Roy Newton for information relating to this type of test. Most of the techniques are in use in other areas of archaeology, therefore I shall describe the type of information that they can provide with respect to glass.

A hand lens, preferably with a flat field of view and large diameter (so that a reasonable area of surface can be examined) will yield very useful information, or a simple microscope can be used if more detail is required. Weathering under various conditions will alter the surface: it may be iridescent, crusted, pitted, crizelled (covered with a network of fine cracks) or merely dull. Wear may show as scuff marks in particular places, often caused by regular use, or there may be surface scratches with associated cracks. Drops of water on the surface which re-appear after the surface is wiped can indicate an imbalance in the constituents leading to alteration of the glass surface (see Chapter 1).

Any process which involves contact with the glass surface during its manufacture will dull its natural fire polish and can give a clue to fabrication techniques. Articles ground from a casting or a solid glass block will usually reveal circular grinding marks, and bottles blown in two- or three-piece moulds will show seams along the paths of the mould joints.

The body of the glass, viewed against a bright light, may be transparent, translucent, or opaque, giving clues as to its composition and the conditions under which it was manufactured. It may be homogeneous or contain seed (small gas bubbles), striae (streaks with a different refractive index from the rest of the glass), cord (thicker inhomogeneities) or stones (opaque inclusions). The seed imprisoned in the rigid glass is the result of chemical reactions occurring during its original melting. The process of removal is known as refining. Nowadays, with better control of temperature and advanced melting techniques, it is possible to obtain bubble-free glass, but old glass often contains many bubbles. Their orientation can sometimes provide information on manufacturing processes. Window glass with seed elongated in parallel rows may have been made by the cylinder process: bubbles in crown glass may lie in arcs of circles, though this is not an infallible guide. Stones in the glass can be unmelted batch materials, bits of the melting crucible which have broken away, accidental inclusions or devitrification products, indicating that the glass was held in a critical temperature range for too long during melting.

The colour of the glass may be a body colour, or it may have been applied as an enamel or stain to the glass surface. Particularly intense colours, such as the deep red produced by copper in glass, were often applied as "flashing", a thin coating layer on a thicker, relatively colourless glass base. Much old glass is a deep green colour: usually this was not intentional but arose as a result of iron as an unavoidable impurity in the raw materials. You should be cautious of glass that is completely free of tint: often water-like clarity is an indication of modern manufacture, but frequent comparison of modern and genuinely old specimens will enable you to distinguish between them.

Low-power microscopic examination extends the studies that can be made with the unaided eye or hand lens. Although magnification need not exceed ×30–60, stereoscopic viewing is essential for looking at contours and spatial relationships. You should also be able to view the specimen by both transmitted and reflected light. The use of polarized light enables you to obtain contrasting views of the specimen and is particularly useful for the examination of cords and striae. Microscopic examination also provides information on manufacturing techniques, especially where these were complex, as in the making of cored vessels and fused mosaic plaques.

Microscopic examination is also very useful for authentification of objects as it is usually possible to distinguish between early glass and modern replacements or fakes, even when a visual examination may be inconclusive. An interesting example is given in a paper by Lanmon *et al.* (1973). The authors were examining a group of "Mutzer" mould-blown glass pieces: this type of glass was originally made in the New England area in the early nineteenth century and was much imitated in the 1920s when prices rose owing to increasing popularity of the ware. The objects in question had been attributed to the earlier period but a series of tests showed that they were in fact fakes, probably made for sale to unsuspecting collectors. Microscopic examination played an important part in coming to this conclusion. Whereas pieces of known early nineteenth century date showed a pattern characteristic of natural wear, with scratches of different lengths and widths, diverging at odd angles, the wear on the Mutzer pieces consisted of a uniformly matt area from which emerged numbers of fine parallel scratches of relatively uniform width and depth. Such "wear" patterns were obviously mechanical, although the difference was not apparent to the naked eye. The body of the glass was also very free from imperfections, inclusions and obvious tint, indicating modern manufacture.

Simple density tests, weighing the object first in air and then in water, can be used to distinguish the denser lead-containing glasses from less dense glass of soda–lime–silica composition. For example, the density of Ravenscroft's lead glass (see Chapter 2) is about 3·15 g cm^{-3}, compared with 2·46 g cm^{-3} for a typical soda–lime–silica glass. Care should be taken with certain Chinese or Japanese glasses, where the high density could be caused by the presence of barium rather than lead. The glass should also be free from gas inclusions such as bubbles, as these lead to false density values. In the case of the Mutzer glasses mentioned above, the densities of the suspected fakes were measured and compared with the densities of a control group. This group was composed of mould-blown objects attributed to a number of different factories and thought to be of early nineteenth century date. Nearly all of the control pieces had densities (and therefore lead oxide concentrations) significantly greater than the Mutzer group. These and other physical, chemical and optical tests established the Mutzer glasses as a distinct group, probably dating from the earlier part of the twentieth century.

One of the series of tests performed on the Mutzer glasses was the measurement of refractive index. If the refractive index of the glass sample is already known, immersion in a liquid with the same refractive index will cause interfering surface features, such as scratches and reflective areas, which hinder observation of the interior to disappear. A range of liquids is available for this purpose: the refractive index of the particular glass will determine the one chosen. By this means it is possible, for example, to distinguish between tiny samples which could be either gem stones or glass: tiny bubbles present in glass and not in precious stones show up when the sample is immersed in toluene.

As described in Chapter 2, medieval window glass is of the potash type, the alkali being supplied from beechwood and other woodland ash. Recent replacements, however, will have been made from soda glass. It is possible to distinguish between the types because the potash glass is weakly radioactive: exposure to a radiation monitoring badge over a period of about two months will cause a slow darkening of the badge.

Standard reference glasses

The previous section had shown how you can learn a great deal about an object by very simple means. To go further requires the use of an increasing number of scientific techniques, which will be described in the following pages. Two excellent review articles by Brill (1969), and by Hench (1975) provide more information on the application of these techniques to glasses, and many references for further reading. Details of many of the techniques themselves are given by Tite (1972), and the references chosen to illustrate the following sections also contain details of useful source papers.

Before these methods can be used with confidence, you must be certain that the results of your analysis are essentially the same as the results that would be obtained by another person analysing the same glass. More generally, this must be the case whenever analyses of different glasses in different laboratories are compared. Otherwise, observed differences could be due either to true compositional differences or to differences inherent in the particular analytical procedures of the laboratories concerned, the accuracy of which depend very much upon the expertise of the analyst. This is true, of course, of any scientific experiment, but glass presents a particular problem because of its complex nature.

In order to see how much agreement could reasonably be expected amongst laboratories analysing the same glass an extensive testing programme was initiated in 1964. Four synthetic glasses were prepared to approximate four types of glass commonly encountered by those working with ancient glasses (Brill, 1972). Two soda–lime–silica compositions duplicated typical ancient Egyptian, Mesopotamian, Roman, Byzantine and Islamic glasses. Glasses containing high percentages of lead and barium approximated to some eastern Asiatic glasses, whilst a potash–lime–silica composition was equivalent to certain medieval and some seventeenth to nineteenth century glasses. Suitable levels of minor and trace elements were also introduced. This was particularly important as it is often only in these elements that meaningful variability occurs. Forty-five laboratories in 15 countries analysed the glasses over a period of several years and only after they had performed the analyses were they given the theoretical compositions. Techniques used included the classical methods of wet chemical analysis (gravimetry, volumetry, flame photometry, atomic absorption spectrometry, colorimetry, coulometry, electrolysis, polarography, and combustion analysis), emission spectroscopy, X-ray fluorescence spectrometry, and neutron activation analysis. The agreement on some elements was quite good, and although there was poor agreement on such elements as calcium, aluminium, lead, barium, boron, phosphorous and chlorine the deviations were not in general greater than might be expected from the theoretical values. For the first time a useful set of standard reference glasses was made available for the routine analysis of ancient glass, making it possible for different laboratories to correlate their earlier results and to calibrate future analyses.

Choice of analytical methods

Given that valid results can be obtained for the complex compositions characteristic of most glasses of interest to the archaeologist, the problem remains as to which method of investigation to select. In practice a combination of techniques is usually chosen. This will depend on many factors, such as specimen size and availability, whether it is desired to examine the surface or the body of the glass, the relative importance of constituents, and so on. It also depends on the purpose of the analysis, which could be the intensive study of a single important object or the compilation of a catalogue of ancient compositions. No one method will completely define a glass as each method has its own advantages and limitations.

Chemical analysis

The classical techniques of wet chemical analysis are particularly useful for the precise determination of major and minor constituents in glasses. They are still extensively used but have the disadvantage of being very time-consuming and needing the skills of a highly trained and meticulous analyst: even so, results from different laboratories can vary considerably. The techniques are destructive and quite large samples are usually required.

Atomic absorption analysis

A technique that is often used because it requires relatively small samples and is possibly the most accurate instrumental method for determining elemental composition is atomic absorption analysis. This technique measures the concentration of elements in the sample, which usually has to be in the form of a solution. The solution is converted into a vapour by a burner unit. Light from a light source, usually a hollow cathode discharge tube containing the element to be determined, is passed through the vapour. The tube is usually designed to emit only the resonance radiation of the element in question, which is then absorbed by the vapour from the sample. The strength of the emerging light signal is then measured and by comparing with the absorbance of a standard solution the concentration of the particular element can be determined. This method is capable of high accuracy and sensitivity for a large number of elements found in ancient glasses, but only one element can be determined at a time, the sample must be in solution, and a single solution cannot always be used for all determinations. Dilutions are required to bring very concentrated constituents into range. Hughes *et al.* (1976) describe in detail the special problems encountered in the application of atomic absorption techniques in archaeology.

Atomic absorption analysis, with its ability to provide quantitative values for major and minor constituents, has proved very useful in the characterization of glass made at a particular site. Brill and Hanson (1976) have used the technique to analyse a number of fragments selected from the site of the New Bremen Glassmanufactory (Frederick County, Maryland, USA) where some of the finest eighteenth century American glassware was produced by a German immigrant, Johann Friedrich Amelung. Brill was able to characterize the chemical compositions of the glass made at the factory and he used his findings to learn something about the way in which the factory operated. For example, although all the glass was of the potash–lime–silica variety, the samples divided into two distinct categories, low CaO content (average 9·05%) and high CaO content (average 19·6%). Visual examination of the fragments showed that these categories corresponded to "fine glass" specimens and to pieces of more ordinary domestic glass respectively, revealing a difference in the basic formulations used in manufacture. Brill speculates that the high-lime content, producing appreciably harder glass, was reserved for objects which had to withstand heavy usage. Such glass was also cheaper to produce as a high-lime composition would have resulted from the use of a low-grade, impure potash. More expensive, purified potash would have been needed for the softer, low-lime composition of the fine glass, which would also be easier to cut and engrave. Of course, the glassmakers of the time, lacking modern chemical knowledge, would have adopted an empirical approach: by trial and error they would develop glasses with particular properties depending upon their choice of original batch materials. In general Brill found that data on major and minor element contents were more useful than data on trace elements for characterizing his glasses.

Atomic absorption has also been used to determine elements responsible for colour and opacity in glass. Lambert and McLaughlin (1978) have analysed Egyptian glass fragments dating from the XVIIIth Dynasty and have shown that the colours resulted from the presence of minor components (typically about 1%) such as copper, iron and manganese: opacity was probably caused by similar minor additions of lead or calcium antimonates. The experiments also showed that the basic glass composition was of the soda–lime–silica type with high concentrations of magnesium and potassium oxides typical of Middle Eastern glass of this period.

Emission spectrography

In this technique a sample of the material under investigation is completely volatilized, usually by an electric arc discharge. This excites the electrons of the atoms of the sample and as they return to their ground state they release energy in the form of ultraviolet, visible or infrared radiation, the radiation consisting of a number of sharply defined wavelengths characteristic of the particular elements excited. Determination of these wavelengths and measurement of the intensity of the radiation at a particular wavelength provides a means of identifying the elements present and estimating their concentrations. The technique has often been used to complement atomic absorption studies: although inherently less accurate it may be more suitable for identifying and estimating trace elements. Hughes *et al.* (1976) used this method for the preliminary analysis of a tiny fragment of the object under investigation, to indicate which elements were present and whether they were major, minor or trace constituents. Because the technique can be used over such a wide range of concentrations it was employed by Sayre and Smith (1974) to analyse a large number of specimens of Egyptian glass ranging in date from the period of the New Kingdom to the early Islamic period. Chronological patterns of composition provided an insight into changes in glass formulation and, in certain cases, geographical origin. Sayre and Smith also found that there were significant correlations between materials used in ancient Egyptian glass and in glass found in other regions of ancient Mediterranean civilization. Conversely, subjecting the results of spectrographic analysis to specially developed statistical techniques, Newton and Renfrew (1970) were able to show that British faience beads were probably local products rather than imports as they formed compositional groups quite distinct from beads from Mediterranean sites, thus helping to cast doubts on the invasionist/diffusionist hypothesis of European prehistory.

Neutron activation analysis

Another technique for determining bulk chemical composition which is particularly suitable for minor and trace elements is neutron activation analysis. Advantages are that it is non-destructive, can be fully automated, is effective for very small samples, and provides rapid sample analysis for a wide range of elements and concentrations: a disadvantage is that access to reactor facilities is required. The specimen is bombarded with slow neutrons which interact with the atomic nucleii of the constituent elements transforming them into unstable radioactive isotopes. These decay emitting gamma rays with sharply defined energies which are characteristic of the particular element excited. A gamma ray spectrometer then measures the intensity of spectral peaks over the appropriate energy range.

The method has proved useful in the characterization of medieval window glass. Olin *et al.* (1972) proved that valid information on composition could be obtained even if the sample available was very small (about 10 mg). The results obtained by neutron activation analysis showed wide variations between different samples, reflecting expected compositional differences resulting from variations in raw materials, melting practices, etc. Olin therefore argued that any samples which showed compositional similarity probably had a common origin. This conclusion has a practical application: restoration pieces in stained glass window panels should be identifiable on the basis of compositional differences because much of the original glass was probably melted in the same workshop (Olin, 1974).

Because this method is capable of high accuracy it has been useful as a discriminating tool in studies of the provenance of faience beads. Aspinall *et al.* (1972) showed that the tin content of British faience beads was significantly greater than that found in groups from elsewhere. This, together with stylistic differences, suggested a different, and perhaps local, source of origin.

X-ray fluorescence analysis

This is a non-destructive technique which is now widely used for the analysis of chemical composition because it is fast, accurate and can be carried out on commercially-available equipment. The specimen under investigation is irradiated with primary, high-energy X-rays which eject electrons from the inner orbits of the constituent atoms. The vacancies in the inner shells are not stable and are filled by electrons from an orbit further out, the energy released in the transitions appearing as secondary or fluorescent X-rays. The wavelength of any X-ray emitted is inversely proportional to the energy difference between the two shells involved in the transition (the higher the energy the shorter the wavelength). Each atom has a characteristic distribution of electron shells and thus each element emits a spectrum of secondary X-rays consisting of peaks at several characteristic and sharply defined wavelengths. The secondary X-rays fall onto a flat crystal plate which diffracts different wavelengths through different angles; by measuring the angles at which the peak signals occur their wavelengths can be determined. The elements may then be identified by the wavelengths of their spectral lines and their concentrations determined from the intensity of the lines (by comparison with intensities for standard samples with known composition).

The technique was previously limited to measurement of bulk concentrations of chemical species (greater than 0·01%) mainly because of inaccuracies caused by absorption of fluorescing X-rays by neighbouring atoms. However, the recent development of modified instruments now makes it possible to measure trace elements in the parts per million range on a routine basis (Hench, 1975), which is invaluable for the study of glass compositions.

Direct analysis is possible without removal of a sample from the specimen. However, X-rays are absorbed strongly by air, the characteristic X-rays from elements with low atomic numbers (which are especially important in the study of glass) being most strongly absorbed. Thus, unless the specimen can be accommodated within an evacuated spectrometer chamber, there is a lower limit of detection of $Z = 22$ (titanium): with an evacuated chamber measurements can be extended down to $Z = 12$ (magnesium) or below. This limits the elements that can be determined in larger objects from which no samples can be taken. In order to obtain a suitable signal-to-background noise ratio a relatively large area of the specimen, typically 1 cm in diameter, must be irradiated: this area must be flat and for many artifacts it is impossible to satisfy the necessary conditions. As several of the difficulties are associated with the use of a diffracting crystal the energy-dispersive (also known as non-dispersive) spectrometer has been developed which dispenses with the crystal. Using a semiconductor detector with an associated electronic amplifier the whole secondary X-ray spectrum is recorded simultaneously. A multichannel pulse height analyser sorts the detector pulses and the data produced are processed by computer equipment according to their energy. The elements present can thus be determined and, by comparison with reference standards, their concentrations measured. A systematic tabulation of spectral data by element and by energy, as well as detailed experimental and computational procedures are given in the paper by Hanson (1973) who used this reference data in his experiments to determine elements in a wide range of museum objects, including glasses. The instrument described in his paper used weak radioactive X-ray emitter sources of low penetration power: thus there were no radiation hazards and the experiment did not cause discolouration of the glass. Also as the sample was not enclosed in a vacuum chamber it was possible to obtain analyses for precious glass objects from which no samples could be taken. Hanson mentions the potential of this instrument in detecting forgeries: trace elements absent in these pieces would be present as impurities in older, genuine articles.

A major difficulty with X-ray fluorescence analysis is that the results obtained represent the composition of a very thin surface layer only (about 0·01 inches deep) because the X-rays are strongly absorbed by matter. This may not necessarily be the same as the bulk composition. Cox and Pollard (1977) stress the importance of sample preparation. They studied a range of glasses of various ages, all apparently in an excellent state of preservation and free from corrosion products. Specimens were examined after washing in acetone, after light polishing and then after grinding off the surface layer. The results showed significant differences, particularly for lighter elements, depending upon the degree of surface preparation, thus showing that the surface had been chemically modified in spite of appearances to the contrary.

The characteristic of surface analysis may be turned to advantage in the study of enamels or stains on the surface of the glass. Brill (1969) describes how the technique was used to determine the nature of decorative stains on samples of Islamic lustre glasses. Comparative analyses of adjacent stained and unstained regions were obtained and were combined with the results of other experiments to place the lustre method within the overall history of glass making.

A development of the technique of X-ray fluorescence analysis which places few restrictions on the size of the specimen and the shape of its surface is the "milliprobe" (Banks and Hall, 1963). In the conventional X-ray spectrometer the diffracting crystal must receive a parallel beam of secondary X-rays, and to achieve this a collimator is used which results in a great reduction in the intensity of radiation to be measured. To obtain final signals of sufficient strength a large area of the sample must therefore be irradiated in the first place and this area must be flat, conditions which are often impossible to achieve. In the milliprobe a very small area of the sample (typically less than 1 mm in diameter) is irradiated and because of this the secondary X-rays originate effectively from a point source. They then fall on a curved (rather than a flat) diffracting crystal and are diffracted to form an image at a point: the detector is located at this point and measures the intensity of the diffracted rays (Fig. 10).

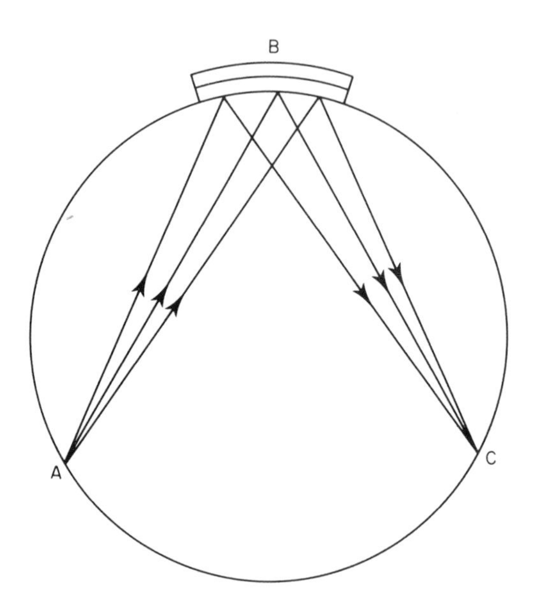

Fig. 10. Schematic diagram of the milliprobe showing positions of source (A), curved diffracting crystal (B), and detector (C).

By moving the crystal and the detector relative to one another and to the sample in a rather complicated way the diffracting angle can be altered and thus the radiation can be resolved into its component wavelengths. Although the total energy of the primary X-rays falling on the specimen is much less than with the standard spectrometer (because the irradiated area is much smaller) the secondary radiation does not have to be formed into a parallel beam with consequent reduction in intensity. The proportion of secondary X-rays detected is therefore very much greater and so a comparable signal-to-noise ratio is achieved. However, as the area analysed is very small, surface unevenness does not seriously affect the accuracy of the results.

The milliprobe has been applied to a range of archaeological problems involving different materials. Hall *et al.* (1964) used it to demonstrate the presence of metallic gold foil as an intermediate layer in poorly preserved beads of the Roman period from Nubia. They were able to show that the beads were of the same formation as well preserved samples found in Northumberland. The comparison would have been difficult without the milliprobe evidence because in cases of advanced decay the irridescence produced in the glass can resemble metallic foil.

Elements with high atomic number, such as gold, are ideal for analysis by the milliprobe. These elements are often also present as constituents of the strongly coloured glasses used for decorative purposes on the surfaces of glass beads. An additional advantage in this case is that the beam of X-rays can be focused on the very small area of decoration rather than on the body. Newton (1971a; 1972a) therefore proposed that the milliprobe be applied to the analysis of decoration on the surface of faience beads. He wished to investigate the possibility that, whilst the body of the beads may well have been made locally from raw materials ready to hand (sand and plant ash), they were decorated by re-melting and applying higher quality glass containing exotic ingredients which had been produced elsewhere. It seemed likely that such specialized coloured glasses, requiring a high degree of knowledge in their preparation, may have been made in only a few centres, and therefore the nature and concentrations of the colouring materials may have become standardized. Thus the decorative glass on the beads might be expected to show similarities in composition. Unfortunately difficulties arise with this type of material because the surface composition measured by the milliprobe could have been altered by weathering or by contamination with foreign elements such as manganese from the soil. Therefore preliminary tests for compositional similarities were carried out on three unaltered Roman glass specimens believed on archaeological and stylistic grounds to be closely related. One specimen had formed part of a moulded bowl. It seemed likely that glass from this bowl had been re-used to fashion other articles, and that the other two specimens were fragments from these articles. General agreement was obtained between the three sets of analyses, and no elements were detected that were not in common. Newton therefore concluded that the milliprobe method could give valuable information on Iron Age beads if care was taken in the choice of beads from different sites.

Electron microprobe analysis

This is a technique that is very similar to energy-dispersive X-ray fluorescence analysis. The specimen is irradiated by a fine electron beam (rather than by X-rays) and secondary X-rays with wavelengths characteristic of the particular element excited are emitted, their intensity being a measure of the concentration of the element. A full account of the technique and its applications is given by Reed (1975). The electron beam is focused on a very small area of the specimen (about 0·2–1 μm in diameter). Thus it is possible to analyse chemically very small selected areas which makes the technique very valuable for special applications such as the non-destructive analysis of miniature mosaic plaques and millefiori glass. Indeed, it is very difficult to analyse this type of material by other means as the taking of even a tiny piece involves the removal of a significant part of the object. Also, as the size of individual features is so small there is a risk of contamination from surrounding areas. Brill (1969) describes some interesting work done using the electron microprobe in the investigation of ancient yellow opaque glasses. It is known that a changeover from an early lead–antimony pigment to a later lead–tin pigment (the opacifying agents) occurred about the fourth century AD. The glass studied by Brill dated from this period. An analysis was carried out on a flake of yellow opacifier which measured no more than 230 μm in its greatest dimension. X-ray fluorescence images showing the distribution of lead and of tin in the flake were congruent, showing that the flake was a lead–tin compound, the later form of the pigment. However, other images produced from the glass revealed the presence of antimony and showed that it occurred separately from the tin in the form of the earlier lead–antimony pigment. Brill concluded from this complex chemical analysis that the glass had been produced by melting together two or more glasses from different periods.

β-ray back scattering

This technique uses the properties of scattered β-rays (electrons) to make quantitative determinations of the amount of lead oxide present in surface layers of glasses or glazes. The technique is described by Emeleus (1960).

When electrons fall on the specimen, some are effectively absorbed whilst others are backscattered and emerge from the specimen surface. The percentage of electrons backscattered depends on the atomic number of the elements present in the surface layer of the specimen, more being scattered backwards from material of high atomic number than from that of low atomic number. If lead ($Z = 82$) is known to be present in a matrix of low atomic number elements, its concentration can be measured by measuring the number of electrons that are backscattered. Calibration is by comparison with samples of glass containing known amounts of lead oxide. The technique is non-destructive and a portable form is available which makes it very useful for museum and archaeological work. The method is only sensitive to lead concentrations greater than about 5%, which means that it can be employed for specimens where lead had been added deliberately, rather than appearing as a trace element.

Auger electron spectroscopy

This is a relatively new non-destructive analytical technique for surface chemical analysis. The specimen is bombarded with a beam of primary electrons which ionizes electron energy levels such as the K-level within the atom. As the atom returns to equilibrium the vacancy in the K-level is filled by an electron dropping from one of the levels further out, such as the M-level. Energy is released and is transferred by the Auger process to an electron in yet another level, further out, such as the N-level. Here electrons are less tightly bound and the energy transferred is sufficient to eject an electron, known as an Auger electron, from the solid. The energy of the Auger electron is characteristic of the particular element. These secondary electrons are collected, measured and their energies analysed. A spectral plot of signal strength against electron energy will show peaks characteristic of particular elements.

The method has been used by Dawson *et al.* (1978) to investigate atmospheric corrosion processes in medieval window glass. Auger electron spectroscopy was considered to be a particularly suitable method as it measures the composition of only the first few atomic layers, and detailed corrosion processes may be influenced by the composition of a surface layer of this thickness, rather than the 10–100 μm commonly analysed by X-ray fluorescence. Also, Auger electron spectroscopy has high sensitivity for elements of low atomic number such as sodium which are important in studies of glass corrosion processes. Unfortunately the electron beam can damage the surface and remove chemical constituents: this affects the signal strengths for various elements, which vary with irradiation time and beam current, and makes the interpretation of results very complex.

Infrared reflection spectroscopy

This is another recent technique that has been used in the study of surface glass corrosion processes. Unlike Auger electron spectroscopy, it does not alter the surface of the glass and it can be applied to contoured glass surfaces of varying area. Analysis is rapid, requiring the use of only standard, inexpensive infrared spectrometers. The incident beam of infrared radiation is reflected from the sample surface and the strength of the reflected signal is plotted over the range of wavelengths of interest (typically 7·0–34 μm). The spectral curves show peaks which are characteristic of the elements present and the chemical bonds that exist between them, i.e. they provide information on the structural units in the glass surface. For example, vitreous silica will show peaks which can be attributed to different types of vibrations of Si–O–Si bonds in the glass network. The addition of network modifiers such as CaO or Na_2O (see Chapter 1) reduces the number of Si–O–Si bonds and weakens their interactions resulting in a change in the peak locations and intensities. Extra peaks are also introduced by the presence of modifying positively charged ions (Na^+, Ca^+, K^+ etc.).

Hench *et al.* (1979) have used the technique to study structural changes associated with aqueous corrosion of simulated medieval glasses. Such studies are particularly important for the conservation of old glass which has often been damaged by such corrosion. During attack of the glass by the water, exchange of hydrogen ions (from the water) and alkali ions (from the glass) occurs and a silica-rich skin forms on the glass surface. Infrared reflection spectroscopy is an ideal technique for monitoring structural changes that occur when the glass has been exposed to water for varying lengths of time and the study of corrosion processes, often very slow at room temperature, can be accelerated by increasing the water temperature.

Analyses of composition and oxidation state

The next two techniques to be described, X-ray photoelectron spectroscopy and spectral transmission, provide information both on composition and on oxidation states of particular elements in glass. These studies are important because the colours of glasses depend upon both factors. You should consult the reference by Bamford (1977) for a detailed explanation of the scientific basis of colour in glass. Briefly, oxides of metals such as iron, manganese, copper and cobalt dissolved in the glass produce the various colours that we see. These metals can commonly exist in more than one oxidation state, and this state is determined by many factors such as the presence of other compounds (which can act as reducing or oxidizing agents), the furnace conditions (the old wood- or coal-fired furnaces often provided a reducing atmosphere) and heat treatment of the glass article. An example is provided by the process of decolourizing glass.

Iron in glass is an intense colouring agent and the removal of this undesirable effect, caused by impurities in raw materials, has always posed a difficult problem for glassmakers. The Venetians from about the fourteenth century added pyrolusite, which contains manganese: this oxidizes the iron and improves the colour of the glass. In the process the manganese itself is reduced, in which form it is colourless. Thus it does not effect the colour of the resultant glass, except by cutting down the amount of light that it will transmit and so reducing its clarity. The reversal of the process can be observed in some old window glass that has been exposed to the sun over many years. The manganese has been oxidized back to its original form and in this state it gives a strong purple colour to the glass.

Thus a careful analysis of both composition and oxidation state can, in principle, tell us a great deal about the conditions under which the glass was manufactured. Unfortunately the situation for ancient glasses is extremely complex, because so many variables are involved (Newton, 1978). Nevertheless such studies increasingly add to our understanding of ancient processes of glassmaking.

X-ray photoelectron spectroscopy

This recently developed technique is now being applied to the study of archaeological materials. A full description is given in the papers of

Lambert and McLaughlin (1976, 1978). The substance to be analysed is bombarded with low energy X-rays which eject electrons from the inner shells of the atom. The binding energy, E_b, of an inner shell electron is characteristic of the specific element involved, being the energy required to eject the electron from the atom. It is given by the expression

$$E_b = h\nu - E_k - \Phi,$$

where $h\nu$ is the energy of the incident X-rays (h is Planck's constant and ν is the frequency of the X-rays), E_k is the kinetic energy of the ejected electron, and Φ is a quantity characteristic of the spectrometer. Thus by measuring the energies of the ejected electrons using an electron spectrometer the binding energies can be determined. Each element in the sample may have several characteristic binding energies (corresponding to the set of inner electron shells). A "photoelectron spectrum" is produced, a plot of the number of electrons ejected as a function of binding energy. This spectrum will show a series of peaks, the positions of which are characteristic of the elements present. The intensity of a peak depends upon the concentration of that element and also upon specific properties of the element and the glass matrix. To obtain quantitative values for element concentrations peak intensities are compared with those of standard samples having a known composition, as in X-ray fluorescence analysis. However, in the case of X-ray photoelectron spectroscopy the standards must match the base matrix even more closely because this method is more sensitive to chemical differences.

Although it is possible to analyse small whole pieces of glass up to 1·5 cm square the surface must be flat and homogeneous to perform an accurate analysis and for larger or more uneven specimens a small sample, usually in powdered form, is needed. Care must be taken that the sample is characteristic of the object as a whole because the ejected electrons come from a very thin surface layer (of the order of 20–50 Å in thickness). Thus this is essentially a surface method, indeed the best available for the analysis of very thin surface features, though great care must be taken to provide a clean surface. It is also possible, by sequential sampling, to obtain analyses that are completely characteristic of any particular level within the sample.

In contrast to X-ray fluorescence analysis, the technique can be used to determine elements with atomic numbers below 10, which makes it very useful for glass analyses. The second great advantage is that elements in different oxidation states have slightly different binding energies for their inner electrons resulting in a different (though related) peak pattern. Taking copper as an example, a sample containing Cu^{+} (the cuprous form) will give a characteristic peak pattern. If the sample contains both Cu^{+} and Cu^{2+} (the cupric form) a second peak pattern will be superimposed. The original peaks will still be there, but extra peaks will appear because of the presence of Cu^{2+}. Lambert and McLaughlin (1976) have used the technique to show that copper was present in samples of turquoise blue Egyptian glass in the cupric state, whilst samples of red Egyptian and Roman glasses contained Cu^{+} or elemental copper. The method can easily differentiate between the copper oxidation states that cause these colour differences.

One drawback of the method is that sample sensitivity is limited. The photoelectrons come only from a very thin layer and the signal strength for elements present in very small concentrations is very low. Thus the technique is not very suitable for the analysis of trace elements, although for major and minor constituents an analysis can be performed with an extremely small sample (as little as 1 μg). Also, because calibration of standards is so difficult, it is not always possible to obtain quantitative data.

Spectral transmission analysis

This technique, unlike the new method of X-ray photoelectron spectroscopy, has been used for routine examination since the 1950s. When light is passed through transparent coloured glass it acts as an optical filter. As the wavelength of the light is varied the amount of light transmitted also varies and the transmission spectrum obtained is characteristic of the metallic oxides present as colouring agents and their oxidation states. As the wavelength is decreased a wavelength is reached at which the percentage of light transmitted drops to zero. This "cut-off" wavelength is directly affected by impurities present in the glass, particularly iron. Lanmon *et al.* (1973) used this characteristic to differentiate between genuine early 19th century glass and pieces which were suspected to be fakes, probably of early 20th century date. The suspect pieces had shorter wavelength cut-offs than the genuine pieces suggesting that they were relatively purer and particularly that they had a lower iron content. This would be expected from the fact that purer raw materials were available in the 20th century than in the

19th. The technique is also very useful for determining the composition of applied surface stains which could otherwise be difficult to investigate. The composition of the base glass can be allowed for by running the spectra against unstained portions of the same glass as a reference standard.

Structural studies

We have seen that the technique of infrared reflection spectroscopy, as well as providing analyses of composition, can give structural information that is of great value to the archaeologist. Two other methods of structural analysis are those of X-ray diffraction and electron microscopy.

X-ray diffraction

When a beam of X-rays is passed through crystalline material diffraction patterns are produced which are characteristic of the crystal phases present. Although glass itself is not crystalline the method has been used for many years for specialized studies of ancient glasses such as the identification of colouring agents and opacifiers. Some of the most important work is summarized by Brill (1969) whose own work is also reported in later papers (1970a, 1973). The method is also useful for identification of inclusions such as unmelted batch materials and devitrification products.

Electron microscopy

The electron microscope may be compared with the ordinary light microscope (with a beam of electrons rather than light illuminating the sample) the advantage being that the electron microscope has much greater resolving power so that microstructural features can be observed.

One of the widely used methods now employed for microstructural characterization of glasses is that of scanning electron microscopy. A range of magnification, 10 times to 25 000 times, and a great depth of field can be obtained, producing detailed pictures of an area of surface with a three-dimensional effect. Werner *et al.* (1975) have used this technique to investigate surface characteristics of ancient glasses affected by weathering, a subject of considerable interest to researchers concerned with the authentification of pieces of unknown provenance.

The scanning electron microscope is often used with an energy-dispersive X-ray fluorescence spectrometer attachment to make a chemical analysis of the observed surface. Rapid determinations are possible and, by comparison with standards of known composition, quantitative measurements can be made. With the addition of an X-ray spectrometer the design of the scanning electron microscope has tended to converge with that of the electron microprobe described in a previous section. The primary purpose of the scanning microscope is to produce high resolution images of the surface, whilst that of the microprobe is to provide analyses at points over the surface, a kind of compositional map, but dual purpose machines are now available which will perform both functions. With such instruments and related developments in electron microscopy the study of chemical composition can be related to that of structure in a very detailed way to provide new information on ancient glasses (Ogilvie *et al.*, 1974).

Isotopic ratios

Whilst the relative abundance of the isotopes of most chemical elements is constant regardless of where they are found it varies significantly, depending upon geological origin, for lead and oxygen and this has important consequences for glass studies (Brill, 1970). The relative isotopic abundances can be determined using a mass spectrometer. Here a beam of ions from the sample under investigation is accelerated by an electric field then passed through a magnetic field where the ions are separated according to their mass. By altering the accelerating voltage ions of different mass, here isotopes of the element in question can be collected at a detector and the ratio of their concentrations determined.

Lead deposits associated with different geological environments can contain significantly different proportions of its four stable isotopes, ^{204}Pb, ^{206}Pb, ^{207}Pb and ^{208}Pb. For example lead ores from Greece, England and Spain, three important ancient mining areas, can be readily distinguished. A study of isotopic ratios for old lead or lead-containing objects can thus provide useful information on the original region from which the lead ore came. Although the method does have disadvantages (see below) it is one of the few techniques giving direct information on geographical origins and it is usually independent of the chemical history of the material (Barnes *et al.*, 1978) being unaffected by such factors as weathering and burial.

Brill (1969) has carried out work on lead extracted from ancient red and yellow opaque glass where it was added as an intentional ingredient. The samples formed parts of panels found in Greece but of a style which strongly suggested Alexandria as the place of manufacture. Brill found, rather unexpectedly, that the leads in the yellow and red glasses were distinctly different from one another and must have come from different sources, suggesting possibly that the glasses were made in different places or produced from earlier glasses from different sources. It was not possible to say anything more specific about origin because ores from widely separated regions can have the same isotopic ratios where they were formed in similar geological environments. Also confusion can arise because leads from different sources can become mixed when metals are salvaged and melted down for re-use.

Brill *et al.* (1974) have also carried out studies on ancient Egyptian glasses where the colourant-opacifier was a lead antimonate, $Pb_2Sb_2O_7$. Yellow opaque glass of this type is frequently found in both Egyptian and Mesopotamian core-formed vessels dating from about 1500 BC onwards, and Brill wished to see if the leads in the $Pb_2Sb_2O_7$ pigments of these two groups were related or not. In these studies supporting evidence from ores and related artifacts would have been very useful, but lead was not abundant in ancient Egypt and so the authors used as comparative material the cosmetic kohl. This was widely used as an eye cosmetic and medication and was often prepared from galena ore. Isotopic analysis of the kohl and yellow Egyptian glass specimens showed that these products constituted a group of leads entirely different from anything found in previous work. Brill therefore concluded that it should be easy to recognize imports into Egypt and "Egyptian" lead in the products of other countries in future studies.

Oxygen isotopic analysis is also useful for glass studies. There are two stable isotopes of interest, ^{16}O and ^{18}O, the heavier ^{18}O isotope occurring naturally in the ratio of about 1/500 with respect to the ^{16}O isotope. However the ^{18}O content, and therefore the $^{18}O/^{16}O$ ratio varies significantly according to the material in which the oxygen occurs, and to its source and this is especially important for the raw materials used in glass manufacture, such as quartz sand and alkali, which typically contain 40–50% by weight of oxygen. Oxygen isotopic analysis thus offers a means for characterizing or grouping early glasses (Brill, 1969, 1970b). One problem is the possibility that variations in melting temperatures and times could affect the ^{18}O content of glasses prepared from the same raw materials, but experiments reported by Brill (1970b) seem to show that the only determining factor is the isotopic composition of the original ingredients.

Dating of glasses

A method for dating glass objects that could be generally applied to yield precise information would be extremely valuable, but no such method exists. The techniques described below can be used in particular circumstances and many sometimes serve to distinguish between ancient and modern specimens even if they cannot give actual dates.

Radiocarbon dating

This technique has been widely used for archaeological dating, particularly in the estimation of age of wooden objects. Living trees acquire small amounts of radioactive ^{14}C from the atmosphere, but this intake ceases when the tree is cut down. By comparing the radioactivity of a modern piece of wood with that of a specimen of unknown age the length of time that has elapsed since the old specimen ceased to live can be estimated. The technique is not in general applicable to glasses which usually contain very little carbon. Also any dates obtained would apply to the raw materials used for glass manufacture rather than to the date of manufacture itself. Brill (1969) reports on the special case of a glass containing exceptional amounts of carbon dioxide, but there has been no widespread application of the technique.

Fission-track dating

This technique can be used with glass and other materials which contain uranium. Its applications are reviewed by Green (1979). The most common isotope of uranium, ^{238}U, can undergo spontaneous fission to form two highly energetic fragments which cause substantial damage to the crystal lattice as they move outwards from the original atom. These fission tracks can be revealed and counted in the glass by acid etching a polished sample and viewing with an optical microscope. The number of tracks depends on the ^{238}U content of the sample and on the time over which the tracks have accumulated: thus this period can be calculated if the amount of ^{238}U is known. To determine this quantity the sample is then irradiated in a nuclear reactor with low energy neutrons which induce fission in the isotope ^{235}U (this forms only about 0·7% of natural uranium). Additional tracks are formed by this neutron-induced fission. A count of these tracks yields a value for the ^{235}U content, and as this bears a constant ratio to the ^{238}U content, the amount of ^{238}U can be calculated.

For a man-made glass the period during which spontaneous tracks have accumulated is usually the time elapsing since manufacture as melting results in the removal of existing tracks. The rate of spontaneous fission is very low and if samples contain very little uranium only a few fission tracks are formed, even over long periods of time. This effectively limits the application of fission track dating as ancient man-made glasses contain, typically, only 0·5–2 ppm of uranium. At these concentrations excessively long periods of observation are required for glass less than 2000 years old: large areas of surface must be scanned in order to count sufficient tracks to give statistically significant results. The possibility of statistical error produces large uncertainties in the results as reported in the work of Yabuki *et al.* (1973) who applied the method to the study of glass vessel fragments from the Karbala desert, Iraq. They obtained ages of 2900 ± 1200 years and 3000 ± 800 years for their samples. However, Nishimura (1971) has obtained fission track ages for glass and glaze samples which agree well with the archaeologically estimated ages: the uranium contents in this case were rather higher than those usually encountered in old glass, being typically 3 ppm.

Natural glass on the other hand is often very suitable for fission track dating: obsidian, widely used for ancient tools and weapons contains uranium up to 20 ppm and objects only a few thousand years old can easily be dated. Wagner (1978) discusses this subject in detail: his paper also surveys the application of the technique to other archaeological materials.

Thermoluminescence

Radioactive impurities or constituents present in a glass produce ionizing radiation which creates free charge carriers, electrons and their positively charged counterparts, holes. Some of these are trapped at defect sites in the glass. If the glass is then heated, electrons are released from their traps and recombine with the holes with the emission of light, the thermoluminescence. When the glass is first melted the stored thermoluminescence in principle is zero and subsequently builds up as the ionizing radiation produces charge carriers. It can be shown that the stored luminescence increases linearly with the age of the glass, and therefore measurement of the luminescence of a sample upon heating should be capable of yielding the age of the glass. Unfortunately exposure to sunlight can significantly alter the amount of stored luminescence: as most glass is transparent or translucent optical bleaching can occur or light-induced thermoluminescence acquired before burial, giving an erroneous date for the glass.

Weathering layers

Glasses subject to corrosive conditions such as burial can develop crusts on their surface where the alkali has been leached out (see Chapter 1). An hydrated residue with a high silica content is left. Some of these crusts have a thickness of up to 4 mm and in cross-section under a microscope can show a continuous sequence of layers from the outer surface to the undecomposed glass. It has been suggested that some type of seasonal variation could result in the formation of one layer per year, thus making it possible, in principle, to date the glass (or rather date the time of burial) by counting the layers. Although there is some evidence to support this view the method could only be used for the very small amount of ancient glass with a sufficiently thick and intact weathered crust (Brill, 1969). More recent work (Newton, 1971b, 1972b) casts doubt upon the "annual layer" theory. Newton considers that the layers may be produced by some physico-chemical process at a fairly constant rate which is generally much less than one layer per year: sometimes the rate could be one per year which may account for the "correct" age of certain samples. Moreover, parallelism of layers would seem to be essential if they were structures laid down annually, and scanning electron microscope studies (Newton, 1972b) show that this is not, in fact, the case.

Hydration rind dating

This technique, which again involves a study of the weathering process has been applied to objects made from natural glass. The freshly chipped or flaked surface on exposure to the atmosphere slowly absorbs water over a period of time to form an hydration layer whose thickness depends upon the time since the fresh surface was exposed and the temperature of the environment. The hydration rind which forms is typically a few microns thick and can be measured microscopically. Barrera and Kirch (1973) have applied the technique to the dating of basaltic-glass artefacts from the Hawaiian Islands chain and find that it gives better results than radiocarbon dating at a fraction of the cost.

Bibliography

Aspinall, A., Warren, S. E., Crummett, J. G., and Newton, R. G. (1972). Neutron activation analysis of faience beads. *Archaeometry* **14**, 27–40.

Bamford, C. R. (1977). *Colour generation and control in glass.* (Elsevier)

Banks, M., and Hall, E. T. (1963). X-ray fluorescent analysis in archaeology: the "milliprobe". *Archaeometry* **6**, 31–36.

Barnes, I. L., Gramlich, J. W., Diaz, M. G., and Brill, R. H. (1978). The possible change of lead isotope ratios in the manufacture of pigments: a fractionation experiment. In: *Archaeological Chemistry 2 . . . Advances in Chemistry Series* **171** (G. F. Carter, ed.), pp. 273–277. (American Chemical Society)

Barrera, W. M., and Kirch, P. V. (1973). Basaltic-glass artefacts from Hawaii: their dating and prehistoric uses. *Journal of the Polynesian Society* **82**(2), 176–187.

Brill, R. H. (1969). The scientific investigation of ancient glasses. In: *Proceedings of the 8th International Congress on Glass, London, July 1967*, Review Lectures, pp. 47–68. (Society of Glass Technology)

Brill, R. H. (1970a). Chemical studies of Islamic luster glass. In: *Scientific methods in medieval archaeology* (R. Berger, ed.), chapter 16, pp. 351–377. (University of California Press)

Brill, R. H. (1970b). Lead and oxygen isotopes in ancient objects. *Philosophical Transactions of the Royal Society of London. A* **269**, 143–164.

Brill, R. H. (1972). A chemical-analytical round robin on four synthetic ancient glasses. In: *Proceedings of the 9th International Congress on Glass, Versailles, Sept.–Oct. 1971: artistic and historical communications*, pp. 93–109. (International Commission on Glass)

Brill, R. H. (1973). Analyses of some finds from the Gnalić wreck. *Journal of Glass Studies* **15**, 93–97.

Brill, R. H., and Hanson, V. F. (1976). Chemical analyses of Amelung glasses. *Journal of Glass Studies* **18**, 215–237.

Brill, R. H., Barnes, I. L., and Adams, B. (1974). Lead isotopes in some ancient Egyptian objects. In: *Recent advances in science and technology of materials* (A. Bishay, ed.), vol. 3, pp. 9–25. (Plenum Press)

Cox, G. A., and Pollard, A. M. (1977). X-ray fluorescence analysis of ancient glass: the importance of sample preparation. *Archaeometry* **19**, 45–54.

Dawson, P. T., Heavens, O. S., and Pollard, A. M. (1978). Glass surface analysis by Auger electron spectroscopy. *Journal of Physics. C: Solid State Physics* **11**, 2183–2193.

Emeleus, V. M. (1960). Beta ray backscattering: a simple method for the quantitative determination of lead oxide in glass, glaze and pottery. *Archaeometry* **3**, 5–9.

Green, P. (1979). Tracking down the past. *New Scientist* **84**(1182), 624–626.

Hall, E. T., Banks, M. S., and Stern, J. M. (1964). Uses of x-ray fluorescent analysis in archaeology. *Archaeometry* **7**, 84–89.

Hanson, V. F. (1973). Quantitative elemental analysis of art objects by energy-dispersive x-ray fluorescence spectroscopy. *Applied Spectroscopy* **27**, 309–334.

Hench, L. L. (1975). Characterization of glass. In: *Characterization of materials in research. Proceedings of the 20th Sagamore Army Materials Research Conference, Raquette Lake, N.Y., Sept. 1973* (J. J. Burke and W. Weiss, eds), chapter 8, pp. 211–251. (Syracuse University Press)

Hench, L. L., Newton, R. G., and Bernstein, S. (1979). Use of infrared spectroscopy in analysis of durability of medieval glasses, with some comments on conservation procedures. *Glass Technology* **20**, 144–148.

Hughes, M. J., Cowell, M. R., and Craddock, P. T. (1976). Atomic absorption techniques in archaeology. *Archaeometry* **18**, 19–37.

Lambert, J. B., and McLaughlin, C. D. (1976). X-ray photoelectron spectroscopy: a new analytical method for the examination of archaeological artifacts. *Archaeometry* **18**, 169–180.

Lambert, J. B., and McLaughlin, C. D. (1978). Analysis of early Egyptian glass by atomic absorption and x-ray photoelectron spectroscopy. In: *Archaeological Chemistry 2 . . . Advances in Chemistry Series* **171**. (G. F. Carter, ed.), pp. 189–199. (American Chemical Society)

Lanmon, J. E., Brill, R. H., and Reilly, G. J. (1973). Some blown "three-mold" suspicions confirmed. *Journal of Glass Studies* **15**, 143–173.

Newton, R. G. (1971a). A preliminary examination of a suggestion that pieces of strongly coloured glass were articles of trade in the Iron Age in Britain. *Archaeometry* **13**, 11–16.

Newton, R. G. (1971b). The enigma of the layered crusts on some weathered glasses, a chronological account of the investigations. *Archaeometry* **13**, 1–9.

Newton, R. G. (1972a). Glass trade routes in the Iron Age? In: *Proceedings of the 9th International Congress on Glass, Versailles, Sept.–Oct. 1971: artistic and historical communications,* pp. 197–205. (International Commission on Glass)

Newton, R. G. (1972b). Stereoscan views of weathering layers on a piece of ancient glass. *Glass Technology* **13**, 54–56.

Newton, R. G. (1978). Colouring agents used by medieval glassmakers. *Glass Technology* **19**, 59–60.

Newton, R. G., and Renfrew, C. (1970). British faience beads reconsidered. *Antiquity* **44**, 199–206.

Nishimura, S. (1971). Fission track dating of archaeological materials from Japan. *Nature* **230**, 242–243.

Ogilvie, R. E., Fisher, R. M., and Young, W. J. (1974). Scanning and high voltage electron microscopy of ancient Egyptian glass. In: *Recent advances in science and technology of materials* (A. Bishay, ed.), vol. 3, pp. 71–84. (Plenum Press)

Olin, J. S. (1974). Neutron activation analytical survey of some intact medieval glass panels and related specimens. In: *Archaeological Chemistry . . . Advances in Chemistry Series* **138**. (C. W. Beck, ed.), pp. 100–123. (American Chemical Society)

Olin, J. S., Thompson, B. A., and Sayre, E. V. (1972). Characterization of medieval window glass by neutron activation analysis. In: *Developments in Applied Spectroscopy* (A. J. Perkins *et al.*, eds), vol. 10, pp. 33–55. (Plenum Press)

Reed, S. J. B. (1975). *Electron microprobe analysis.* (Cambridge University Press)

Sayre, E. V., and Smith, R. W. (1974). Analytical studies of ancient Egyptian glass. In: *Recent advances in science and technology of materials* (A. Bishay, ed.), vol. 3, pp. 47–70. (Plenum Press)

Tite, M. S. (1972). *Methods of physical examination in archaeology.* (Seminar Press)

Wagner, G. A. (1978). Archaeological applications of fission-track dating. *Nuclear Track Detection* **2** (1), 51–64.

Werner, A. E., Bimson, M., and Meeks, N. D. (1975). The use of replica techniques and the scanning electron microscope in the study of ancient glass. *Journal of Glass Studies* **17**, 158–160.

Yabuki, H., Yabuki, S., and Shima, M. (1973). Fission track dating of man-made glasses from Ali Tar Cavern vestiges. *Scientific Papers of the Institute of Physical and Chemical Research* **67** (1), 41–42.

4

Glass Compositions and Raw Materials

Introduction

The complexity of ancient glass compositions, arising from the variety of batch materials that were used makes it very difficult to draw conclusions about the source of these materials. Common glass of soda–lime–silica or potash–lime–silica composition is made from ingredients that are usually readily available, and given a plentiful supply of fuel, coal or wood, and clay for melting pots it was not too difficult to set up a glasshouse: glassmaking sites are therefore widely distributed, but obviously some places were more favoured than others. The reasons for this will be discussed in more detail in Chapter 6, but here we shall look at the raw materials themselves to see what information they can give on glass manufacture. In this connection we have to remember that we are speaking with the benefit of scientific hindsight: the knowledge of the old glassmakers was purely empirical and glass, with its endless possibilities for variation, must have raised many questions in their minds (as indeed it still does). Great confusion frequently arose over the identification of raw materials: for example sodium and potassium were not differentiated until the late seventeenth century, and it is very difficult to follow many of the ingredients and directions in the old texts because the same word is used for different substances, and vice versa. However, successful glass compositions were eventually arrived at by trial and eror over a long period and careful selection from amongst the raw materials that were available. These compositions, at least with respect to the major constituents were not too different from those of modern glasses, for the range of mixtures which makes satisfactory glass is not very large. Sand, as a source of silica, is the first requisite and to this must be added an alkali, soda or potash, to lower the melting point and make the glass more easily fusible. It is, in fact, possible to lower the melting point below 800°C but a glass with this alkali content would be highly susceptible to attack by water and would not be a practical material. Lime, added unintentionally in the early days (see below) increases the durability but too much lime gives a glass prone to crystallization and of poor durability.

Nearly all recipes for making glass from 700 BC to the seventeenth century prescribed crushed silica rock or sand, and ash (or "glassmakers' salts") as the major constituents. Yet nearly all analyses disclose the presence of from 2 to 3 up to more than 20% of lime, 0·2–7% of magnesia, and small amounts of alumina and other oxides. These constituents must therefore have come from the major batch materials or from the corrosion of the crucible. This argument is developed in greater detail by Turner (1956c). Professor Turner was one of the first persons to look at ancient glasses in a truly scientific manner: his studies remain a model of logical reasoning backed up by proper evidence which should be read by all those interested in the history of glass. His paper, written in 1956, was the fifth in a series on ancient glasses and glassmaking processes: others are cited in the bibliography at the end of this chapter (Turner, 1954, 1956a,b).

Sands

The two sands for glassmaking which are specifically mentioned by location in classical writings are those at the mouth of the River Belus on the Syrian coast and the seashore deposit mentioned by Pliny near the mouth of the river Volturnus, north-west of the ancient harbour of Pozzuoli and of Naples. The Belus sand was highly reputed over many centuries and is mentioned as being used for glassmaking by Strabo in the first century BC and by Pliny, Josephus, Tacitus and others in the first century AD. Turner found that it contained lime equivalent to 14·5–18% of calcium carbonate, 3·6–5·3% of alumina and about 1·5% of magnesium carbonate: when mixed with alkali it would have been quite possible to make a durable glass. The colour of the glass produced would not have been up to the standards of modern colourless glass, because although the iron oxide content was low in comparison with other sands that Turner examined it was high in comparison with that of present day glass.

The sources of sand employed in ancient Egypt are unknown. Most Egyptian glasses contain substantial amounts of iron and Turner considered that only impure sands such as the sands of the desert could have supplied such quantities. He supported these conclusions by a series of analyses which also confirmed the necessary substantial lime contents: the lime could have been incorporated by wind-driven material scouring limestone bluffs. It is interesting to note that similar calcareous sands are present on British coasts, for example at Hayle in Cornwall and at Elie on the coast of Fife. Both were tried for glassmaking in the early 1920s but the lime concentration must have been too high, because in both cases the glass in the furnace underwent extensive devitrification. It is quite possible that the ancient glassmakers had the same trouble, in which case they would have had to add quantities of much purer silica sand.

Documentary evidence exists for sources of glassmaking sand in England. Christopher Merrett, writing in 1662, distinguishes between the types:

> Our glass houses in London have a very fine white sand (the very same that's used for sand-boxes and scouring) from Maidstone in Kent, and for green glasses, a coarser from Woolwich. . . . Both these cost little besides their bringing by water.

Washed sand from Lynn, mentioned in eighteenth century glass recipes, is still used on account of its high purity, and another valuable source of sand was Alum Bay in the Isle of Wight. Glassmakers also used crushed quartz rock as a source of silica, an idea that may have been imported from Italy where "sparkling white river stones that are clear and breakable" were recommended by a sixteenth century glassmaker.

Sources of alkali

Alkalis for glassmaking have been obtained from many different sources in the past and the compositional complexities of the raw materials make it very difficult to define places or substances of origin.

Up until the medieval period both in Western Europe and the East the dominant alkali in ancient glasses was soda. Sources of alkali available to the glassmakers included natural deposits resulting from evaporation and drying up of land-locked seas and lakes, and salts obtained by deliberate evaporation of sea or river water in pans or pits. It is extremely likely that natron (a compound containing sodium carbonate and bicarbonate) from the Wadi Natrûn (an ancient glassmaking site to the north west of Cairo) was used for glassmaking: it had been employed from very early times as a detergent, in medicine and for embalming. The composition of the natron is complex and variable: the sodium carbonate content typically varies from 22·4–75·0%, sodium bicarbonate from 5·0–32·4%, sodium chloride from 2·2–26·8%, sodium sulphate from 2·3–29·9%, as well as water and insoluble material. However, Egyptian glasses do contain potash although it is not the dominant alkali, and analyses reveal more potash than could be derived from natron or natural soda. Also it would have been necessary to use a calcareous sand to provide the necessary lime for stability.

Evaporation from lakes or rivers also produces salts of varying compositions. Pliny, writing in the first century AD, said that this process was carried out at Naucratis in the Nile delta and at Memphis (south of modern Cairo) using the Nile waters which were passed through sluices and channels into shallow pits where the evaporation yielded "salt".

A nineteenth century analysis of salts in solution in the Nile from a point south of Cairo revealed a complex composition: sodium, calcium and magnesium carbonates were present as major constituents, with potassium carbonate and several other salts as minor constituents. As seven thousand parts of water would have to be evaporated to yield one part of salt the work involved would be considerable, but natural enrichment of the sands of the Nile would occur after the annual flooding when the waters had subsided and evaporated away.

The other major source of alkali was plant ash. This could produce a soda- or potash-type glass depending on the plant composition. We know that the Assyrians in the seventh century BC were using this substance for glassmaking. Their recipes and processes are recorded on clay tablets from the Royal Library of Assur-bani-pal. The alkali source was salicornia, a soda-containing plant. Although it was originally thought that the tablets mentioned lime as a deliberately added ingredient, new translations which take technological considerations into account show that this is not so: lime would always be introduced accidentally, in quantities sufficient to give a durable glass.

The question of lime is an interesting one. A deficiency in lime (typically a content of less than about 4% by weight) renders a glass liable to corrosive attack by moisture. A stable glass, if it does not contain too much alkali, will contain typically 7–10% of lime. Turner (1956c) shows very convincingly by means of analyses that lime, a major constituent of nearly all ancient glasses, was added as a constituent of the sand and/or alkali rather than deliberately. The raw materials also yielded considerable alumina, varying amounts of iron oxide and magnesia as well as other constituents found in these complex glasses.

To return to plant ash, this could provide sodium and potassium carbonates, much chloride and sulphate, calcium and magnesium carbonates and phosphates; like other alkali sources a raw material of complex and widely varying composition. This variation is not confined to that between different types of plant (for example, between a coastal plant relatively rich in soda and an inland plant relatively rich in potash). Plants draw their mineral content from the soil, and if this becomes impoverished, or its composition changed, then the mineral content of plants grown in that location will also change. Thus ash from any particular plant, for example, beechwood ash, will have a different composition according to the locality in which it is grown, making it even more difficult to form a link between the glass and its original raw materials. It must also have added considerably to the problems of the old glassmakers in their attempts to produce usable glass. This is why glassmaking texts provide such differing recipes for glassmaking. For example, the twelfth century German monk Theophilus prescribed

> two parts of the ashes of which we have spoken (beechwood) and a third of sand carefully purified from earth and stones, which sand ye shall have taken out of water . . .

Biringuccio in 1540 described a similar mixture but of different proportions:

> The method of composing glass . . . First one takes ashes from the saltwort of Syria . . . Now some say that this ash is made from fern and some from lichen . . . [Then take] some of those sparkling white river stones . . . (or) a certain white mine sand . . . two parts are put to one of the said salt and a certain quantity of manganese according to your discretion.

However, as explained in Chapter 1, it is possible to prepare glasses with a relatively wide range of compositions (in comparison with chemical compounds where the constituent parts combine in fixed amounts) so the glassmakers must have proceeded by way of many trials until they arrived at compositions which worked for them, even though there was an ever present risk of failure through unintentional variations in composition of raw materials or melting processes. This risk persisted until modern times, when it became possible to exercise precise control over melting and materials of known composition were available having the required degree of purity. Mistakes can still occur: cullet (waste glass introduced as part of the batch materials) of the wrong composition, new working furnaces and failure to appreciate the complexities of the melting process are examples of factors which can cause trouble for present day glassmakers.

At some time prior to 1000 AD a great change came about in the type of glass produced in Western Europe north of the Alps: analyses of medieval glass of this period show an almost complete change to potash as the major alkali. The source of this potash varied: beechwood was widely available and was the ash mentioned by Theophilus:

> If you should decide to make glass, first cut plenty of beechwood logs and dry them. Then burn them together in a clean spot and carefully collect the ashes, taking care not to mix in any earth or stone.

Newton (1980) speculates that this swing coincided with the growth of demand for window glass for churches and cathedrals after 1000 AD which created a shortage of the marine plants from coastal sites. There were extensive beechwoods in continental Europe at this time and the glassmakers used the wood for their furnaces and its ash as the source of alkali, as described by Theophilus and several other writers. Other sources could, of course, be used and the determining factor would often be local availability. Wood (1981) has surveyed the Surrey–Sussex Wealden sites and has shown that the glass houses were almost certainly sited in oakwoods. Although beech does occur, along with chestnut, ash and sycamore, the woodlands are essentially oak.

Much work remains to be done on this subject but it is interesting to note that there is a very old reference to oak as the source of alkali: Pliny in referring to the production of "salt" states that the ash of the oak (*quercus*) was no longer in use in his day. Documentary evidence also exists for other plants, including ferns, pinewood, cods and stalks of beans, bramble berry bushes, millet stalks, rushes, reeds and thistles.

As Newton points out the change to beechwood and other inland plant sources resulted in an exceedingly high lime–potash glass with such poor durability that much has vanished and the rest is now in great peril. There was also a positive side: large and variable amounts of iron and manganese in beechwood ash made it possible to obtain glass in a wide range of colours by varying the melting conditions in the furnace.

We know from Antonio Neri's *L'Arte Vetraria* of 1612 that there was also a trade in sources of alkali which presumably had been carried on since ancient times. He wrote of polverine or rochetta, the plant ash exported from the Levant, Syria and Egypt. His translator and commentator, Christopher Merrett, writing in 1662 mentioned four types of plant from which the Alexandrians prepared polverine, namely *kali geniculatum, kali secunda, kali Egyptiacum* and *kali spinosum*. He identified *kali spinosum* from a specimen which happened to be in a bag of imported polverine and he stated that the species also grew on British coasts and on the banks of the Thames, but that British plants yielded very little ash when compared with those from the Levant.

Another source of alkali at this time (and until the early nineteenth century) was barilla. It was prepared from coastal plants near Alicante in Spain and was widely exported from the end of the sixteenth century. Neri expressed a preference for the use of polverine and rochetta over barilla (which seems to have been exceptionally rich in alkali). Turner (1956c) was able to obtain further information on the type of plant which might have been one of the sources for polverine. In 1943 a review was undertaken of resources in Syria with a view to developing new or existing industries. In the course of this work a study was made of a product known locally as "keli", the residuary ash from the Syrian desert plant, chinane. The dried plant was said to yield one quarter of its weight in ash and some 50 different samples were analysed, giving an average sodium carbonate content of 45%.

Seaweed from the western shores of the British Isles and the coasts of France was also burnt to produce an impure alkali for glassmaking which was known as kelp in Britain and *varec* in France. This source became increasingly important, and by the end of the eighteenth century "kelping" was a major industry, only declining in importance as manufactured soda became widely available. The history and economics of the kelp industry are described in detail by Chapman (1970).

Variations in batch composition and problems of glass melting

Alkali available to the ancient glassmaker from all sources, as well as being extremely variable in composition, contained a high percentage of material that was not readily reactive: from 5–80% of the raw materials could not be incorporated into the glass. Turner (1956c) discusses the melting reactions in detail, but the old glassmakers also recognized the variability in quality of the ash that they had to use. Neri discusses the problems of making and reproducing workable glass and it is worth looking at his recommendations in detail for the light that it throws on past glassmaking practices.

Impurities and undesired constituents were removed by a series of operations. The raw materials were carefully selected and the ash was then purified. Crude ash, polverine, was crushed finely and boiled repeatedly with successive quantities of water in brass cauldrons, the extracts being ladled out into earthenware bowls to settle, and the clear liquid evaporated until crystallization occurred. Before boiling, 12 lb of partly calcined tartar (potassium hydrogen tartrate) were added to each cauldron: this would have acted as a reducing agent and was said to produce "more and whiter salt". Crystals of this salt were then skimmed off, dried, heated till hard, crushed and sieved. Unfortunately, although this process produced a purified ash which gave a glass of more brilliant appearance it also also removed stabilizing calcium and magnesium carbonates and phosphates and the resulting glass was not very durable. The seventeenth century glassmakers did not understand the chemical processes involved but Merrett does say that:

> in the finest glasses, wherein the salt is most purified, and in a greater proportion of salt to the sand, you shall find that such glasses standing long in subterraneous and moist places will fall to pieces, the union of the salt and sand decaying.

Following purification the ash and the "tarso" (powdered silica rock or sand) were then weighed, mixed and heated in two stages. In the first the mixtures were raked together for several hours on the hearth of a reverberatory furnace called a calcar. (In such a furnace the material being treated, in this case the mixture of sand and ash, is heated indirectly: the roof is heated by flames and the heat is radiated down onto the material off the roof.) The temperature was too low to produce complete fusion and liquefaction but high enough to bring about the initial stages of melting, burn off some impurities and produce a granular substance known as frit or fritt. Neri says that "the fritt thus made becomes as white as snow from Heaven." The frit was then placed in pots with a quantity of manganese (to act as a decolourizer) and the first melting took place. The product was cast into large earthenware pans full of clean water which had the effect of removing from it "a sort of salt called sandever, which hurteth the Crystall, and maketh it obscure and cloudy." This process was repeated until all the sandever had been removed. Merrett refers to this impurity as *suin de verre* or fat of glass and says that it was taken out by skimming with a ladle as it floated on top of the molten glass. Sometimes the amount was so great that fresh ash had to be added four or five times before the pot was full of good glass. The sandever was not wasted, however:

> tis sold into France, and there used to powder their meat, and to eat, instead of common salt; a solution hereof bestowed on garden-walks destroys both weeds and vermin.

Both Neri and Merrett were insistent that the sandever be removed, otherwise it made the glass unfit for working, very brittle and "in no way plyable."

Before leaving this discussion of problems caused by the complex composition of the alkali I shall stress again the difficulty of assigning sources to the alkalis used in ancient glasses. Soda is predominant over potash in the old glass of all periods except of medieval and renaissance times in Western Europe north of the Alps, following the adoption of beechwood and other inland plants. Whilst plants grown near the sea or on salt desert land have a relatively high soda content and those inland are relatively rich in potash there is not a sufficiently clear-cut difference between them which enables us to say what was the source of alkali used in a particular glass. For example analysis of samples of the ash from seaweed, *varec,* reported by Turner showed that it contained 31% of potassium compounds and 32% of sodium compounds. Furthermore, the final glass composition would depend not only on the original total composition of the alkali materials but also on the amount of each capable of reaction with the sand during the melting process.

Colour in glass

The history of colour in glass is fascinating but very complex. Because of lack of chemical knowledge the old glassmakers could not identify colouring agents uniquely. They therefore used the same name for different substances and vice versa which in turn confuses us when we read their accounts. The correlation and interpretation of information contained in glass recipes and accounts over the centuries would be an interesting task but it is one that still has to be carried out. Therefore I shall not concentrate here on the fine details involved in the production of a particular colour from a specific raw material (see Chapter 2 for some of the commonly used colouring agents and their sources) but look at some aspects of colour in the context of the development of glassmaking.

There has always been a certain air of mystery about the production of colour in glass. Recipes and processes were closely guarded secrets at least up the nineteenth century and were handed down from generation to generation. (As the glasses could not be analysed the makers were reasonably safe unless one of their number left taking the recipes with him.) There are still some people who believe the old story that gold ruby glass can be produced by throwing a gold coin or ring into the molten glass, although scientific evidence shows that this is impossible. The rather difficult process of producing the rich ruby colour, which involves a reheating of the glass, could be made to look as though it depended on the use of a coin if this was thrown in at a carefully determined time (and, no doubt, retrieved afterwards). The mystery also helped to hide the very real uncertainty about the finished product. Neri's book is full of interesting recipes for all sorts of coloured glasses but one wonders how often they were attained in practice. That skill at all stages was vital is made clear in Merrett's commentary on the production of coloured glass:

> Wherefore (he) always puts in all his colours, not by weight, nor measure, but by little and little at a time, and then at each time mixeth them well with the metall, and taketh out a proof, and by his eye alone judgeth whether the colour be high enough, and when too low adds more of them till he attain the desired colour. . . . some thereof is good for little, other very rich, some of a middle nature, and to know their difference in goodness, there's no way found out but tryall in the furnace.

You should therefore beware of coming to conclusions about say, the selection of colours for a stained glass window. We tend to think in modern terms, because glass of a standard colour is nearly always available. The old selection processes must have been much wider than we think. Newton (1978) has written an interesting paper on this subject, with particular reference to manganese in medieval glass. He mentions experimental meltings by Sellner and co-workers (1979) of medieval-type glasses containing only manganese and iron oxide as the colouring agents. By changing the furnace atmosphere (over which the glassmakers would originally have had very little control) she was able to produce a wide range of colours, from bright blue when the atmosphere was fully reducing, through green and amber to brownish purple when the atmosphere was fully oxidizing. The old glassmakers took advantage of what, to them, must have been somewhat arbitrary colour changes: Theophilus says:

> If you perceive that the contents of any container happen to change to a tawny colour, which is like flesh, let this glass serve for your flesh colour. Take as much of it as you want. Heat the rest for two hours . . . and you will obtain a light purple. Heat again from the third to the sixth hour and it will be a perfect reddish purple.

As Newton points out the colours obtained by the medieval glassmakers were determined by many factors such as the purity of their raw materials, the times and temperatures of melting, the reducing or oxidizing furnace atmospheres, over which they had little control, and therefore they must often have selected from available colours rather than being able to determine the colours in advance. That this situation persisted for a very long time is borne out by the evidence of two remarkable notebooks in the possession of the Frank Wood Joint Library of Glass Technology at the University of Sheffield. On careful examination they proved to be the record of an unbroken series of experiments, concerned with the production of clear and coloured glasses, covering the years 1778 to 1802. It seems likely that they were the record of a Midlands glassmaker, who must have been a most methodical and painstaking worker, for he describes hundreds of trial meltings, varying composition and melting conditions and carefully noting the results. His approach, however, was still empirical. The following account is typical (he was experimenting with a yellow glass for windows):

> I have not put any magnees to this batch because the metal of the last pot had too large a quantity in it, which made it very much on the brown cast, it was also too deep and therefore I put to this only 3 lb of colour to a score (viz. iron) instead of 4 lb which was used last week . . . I have added to this 1/8th of the weight of colour of arsnike (than used before), which I apprehend will not be too much if it will make the yellow colour brighter and more of a true yellow. By a proof on Sunday morning the colour appeared to be rather too deep . . . unless the muffs were blow thin . . . when they were a very good yellow. (Muffs were cylinders of glass.)

The late eighteenth century had seen a quickening in the pace of scientific discovery. The difference between soda and potash was finally made clear, methods of chemical analysis were developed, and new elements such as uranium were discovered which could be used for the colouring of glasses. It now became clear that the old names such as crocus martis, tartar and ferretto were both confusing and inexact, and a systematic nomenclature in which symbols were given to various compounds was introduced in 1787. Perhaps the greatest influence on glass technology was that of J. J. Berzelius who from 1808 onwards laid the foundations of quantitative chemical analysis and showed how the equivalents and atomic weights of elements could be used to make quantitative chemical predictions.

Changes soon took place in the glass industry. The new chemical knowledge was quickly applied to glassmaking processes and although throughout the nineteenth century accounts of glass manufacture show varying degrees of scientific awareness it is appropriate to end this section with a quotation from the *Chemical Catechism* written in 1826 by S. Parkes:

> The art . . . of making glass is . . . entirely chemical, consisting in the fusion of siliceous earth with alkali . . . (The manufacturer) will be enabled on chemical principles to ascertain the exact quantity necessary for any fixed portion of silica. . . . Metallic oxides and oxides of lead make glass more fusible. Glass cannot be made without great heat for it is one of the laws of nature that in order that two bodies may become chemically united, one of them must be in a state of fluidity.

Lead in glass

As mentioned in the above paragraph, lead lowers the melting temperature of glass, and its use, for a variety of purposes, has a very long history and is therefore of interest in archaeological studies: it has also been the subject of extensive scientific analysis and has great importance for the history of British glassmaking.

Lead seems to have been mentioned as a constituent for glazes as far back as 1700 BC on an inscribed set of Babylonian clay tablets, and a very early piece of glass with a substantial lead content is a blue fragment from Nippur in Mesopotamia dating from about 1400 BC. Lead is also apparently mentioned in the seventh century BC on Assyrian clay tablets from the Royal Library at Nineveh. Circular cakes of sealing-wax red glass containing lead of approximately the same period found at Nimrud support the view that this specialized glass was an article of trade in ancient times, perhaps being used as an inlay for metals (a use that has continued until recent times). Theophilus in the twelfth century AD was certainly familiar with lead as a constituent of glasses for he describes a way to make glass rings, the batch for which contained lead. Somewhat later Heraclius, in his treatise *De Coloribus et Artibus Romanorum* says:

> Take good and shining lead, and put it into a new jar and burn it in the fire until it is reduced to powder . . . Afterwards take sand and mix it well with that powder, but so that two parts may be of lead and the third of sand, and put it into an earthen vase (crucible). Then do as before directed for making glass, and put that earthen vase into the furnace and keep stirring it until it is converted into glass.

Although lead-containing glasses have been found in many places, dating from ancient and medieval times, they were essentially substitutes for gemstones or the basis for enamels. A substantial part of Neri's book (1612) is concerned with the making of lead glass and he describes the production of many beautiful colours: emerald, sapphire, lapis lazuli, topaz, ruby, opal and garnet:

> These pieces of crystall may be wrought as a jewel at the wheel, and will receive a good polishing, lustre and shewing beauty . . . they make a fair shew, being set in gold.

However, Neri also describes the production of the basic "glass of lead" from calcined lead and crystal frit (the alkali-silica frit described in a previous section) and says that it can be made into drinking or other vessels. This may have been an influence on George Ravenscroft when he started to experiment with the production of lead glass for vessels towards the end of the seventeenth century (see Chapter 2).

Neri's glass was difficult to form: when it was rolled on a marble surface to shape it the marble had first to be wetted, otherwise it stuck to the glass surface and spoilt it. Merrett in his comments of 1662 says that:

> Glass of lead, 'tis a thing unpractised by our Furnaces, and the reason is, because of the exceeding brittleness thereof. . . And could this glass be made as tough as that of Crystalline, 'twould far surpass it in the glory and beauty of its colours.

Merrett had, however, performed, or seen performed many experiments on the production of lead glass of various colours and he describes problems such as the tendency of the lead to separate and break through the bottom of the pot, so it seems likely that Ravenscroft was continuing work that had been set in train earlier in the century. Moreover his were not the only experiments on the production of high quality glass: evidence in state papers of 1666 shows that the Duke of Buckingham, a leading manufacturer and experimenter, was using saltpetre (potassium nitrate) as a flux in the manufacture of fine glass. As described in Chapter 2 Ravenscroft also had considerable difficulties with his glass: the first attempts were readily attacked by water because too large an amount of alkali was needed to flux the silica. When he added lead oxide he reduced the melting point and so required less alkali: he could then produce a stable glass having a clarity and brilliance unknown up to that time. Although lead oxide is itself a glass former and glass samples containing up to 90% are known, it is believed (from the analysis of a piece reasonably attributed to Ravenscroft) that his glasses contained approximately 15% of lead oxide. In general the content varied from about 30–35% from the late seventeenth century to the present day: this proportion in combination with potash gives the most brilliant, stable and workable glass.

The popularity of lead crystal spread rapidly and a very large proportion of English glasses from this time are made from lead glass. However, as it was quickly taken up on the Continent it is not correct to assume that any samples you may find are necessarily of British manufacture.

The lead glasses of Ravenscroft (as well as many others) still cause many problems owing to their instability and it has been shown that a major factor is a deficiency of lime, presumably removed from the original alkali batch material by the purification process. It is ironical to think that the very efforts made by the workers to ensure that their glasses had the best possible appearance should also bring about their decay and eventual destruction. Newton (1980) refers to studies by R. H. Brill on the design of museum cases to combat this particular problem. Cases tend to become too wet in summer and too dry in winter, but atmospheric control can be achieved by introducing a large amount (about 6 kg per cubic metre) of "equilibrated" silica gel, made by wetting the gel with about 25% of its dry weight of distilled water: it is then possible to stabilize the relative humidity at about 45%. Nor is this a problem that has been completely overcome: I have in my possession some modern Scandinavian lead crystal glasses which develop an alkaline surface bloom unless frequently washed. Such instability can be misleading, sometimes giving a spurious appearance of age to quite recent glass.

For further reading on lead glass you should consult the review article by R. J. Charleston (1960). Turner (1956a) deals with the subject in a general survey of the chronology of glassmaking constituents, and Biek and Bayley (1980) review recent developments, especially finds from excavated English sites. They also survey the use of other glass constituents.

Bibliography

Biek, L., and Bayley, J. (1980). Glass and other vitreous materials. *World Archaeology* **11**, 1–25.

Chapman, V. J. (1970). The early kelp industry and iodine production. In: *Seaweeds and their uses*. 2nd ed. (Methuen), chapter 2, pp. 24T–39T.

Charleston, R. J. (1960). Lead in glass. *Archaeometry* **3**, 1–4.

Newton, R. G. (1978). Colouring agents used by medieval glassmakers. *Glass Technology* **19**, 59–60.

Newton, R. G. (1980). Recent views on ancient glasses. *Glass Technology* **21**, 173–183.

Sellner, C., *et al*. (1979). Investigation of the relation between composition, colour and melting atmosphere in ancient glasses (forest glasses) by electron spectroscopy and electron spin resonance (ESR). *Glastechnische Berichte* **52**, 255–264.

Turner, W. E. S. (1954). Studies of ancient glasses and glass-making processes. Part 2. The composition, weathering characteristics and historical significance of some Assyrian glasses of the eighth to sixth centuries B.C. from Nimrud. *Transactions of the Society of Glass Technology* **38**, 445T–456T.

Turner, W. E. S. (1956a). Studies in ancient glasses and glassmaking processes. Part 3. The chronology of the glassmaking constituents. *Transactions of the Society of Glass Technology* **40**, 39T–52T.

Turner, W. E. S. (1956b). Studies in ancient glasses and glassmaking processes. Part 4. The chemical composition of ancient glasses. *Transactions of the Society of Glass Technology* **40**, 162T–186T.

Turner, W. E. S. (1956c). Studies in ancient glasses and glassmaking processes. Part 5. Raw materials and melting processes. *Transactions of the Society of Glass Technology* **40**, 277T–300T.

Wood, E. S. (1981). Private communication on the woodlands of the Wealden glassmaking sites.

5

The Conservation of Glass

Introduction

Chapter 3 has shown how the scientific study of glass can yield information of value to the archaeologist in his or her study of the past. Such studies are also vital in the field of glass conservation, because they provide us with the basic information on which to base sound conservation practices. A great deal of work is now undertaken, in many countries, which is aimed at understanding decay processes and developing techniques which will arrest or retard them without damage to the original glass.

It might be thought that because of these efforts a set of guidelines would by now have emerged which could be followed by archaeologists when dealing with any glass object found during the course of excavation or other exploratory programme. Unfortunately the situation is not so straightforward. Virtually every technique which has been proposed for conservation has its supporters and opponents, who frequently hold widely differing views, often based on actual studies which have yielded contradictory results. The literature of glass conservation reveals a scientific minefield, where cherished opinions are likely to be exploded at any time by further work. More seriously, perhaps, there is still a great deal of uninformed speculation upon causes of glass decay which is not based upon a proper understanding of the nature of glass, but it is to be hoped that this will become less of a problem with the advance of co-operative scientific work.

In Chapter 1 I discussed the factors which affect the durability of glass but pointed out that the actual mechanisms involved are still a matter of considerable discussion. In view of this fact any information of an empirical nature may well be capable of yielding practical help in the urgent problems of glass conservation if the work is carefully carried out. With this aim in mind experiments are being conducted by the Experimental Earthworks Committee of the British Association for the Advancement of Science and others on the burial of technological materials, including glass, in artificial earthworks in order to give information about rates of deterioration of these materials. Newton (1981) gives a summary of progress of the experiments.

Nine glass samples have been buried at Wareham Heath, Dorset, and glasses with nominally the same composition at Ballidon, Derbyshire: these environments are acid and alkaline respectively. The compositions reproduce typical Roman, medieval, Saxon and several modern glass formulations. The Ballidon burial is possibly of most interest, because the alkaline environment would be expected to have the greatest effect on the samples: the first samples were buried in 1970 and there is a detailed timetable for re-excavation until the year 2482. Whether this will prove to be possible only time will tell, but all is going well at the moment, the chief difficulty being the displacement of samples by rabbit holes. So far only the simulated Saxon glass buried in alkaline soil has shown signs of extensive deterioration, although it might have been expected that the simulated medieval glass, containing nearly 17% K_2O and 19% CaO, would have deteriorated in a similar manner.

If we look into the problem a little further, we should not be surprised that glass deterioration is such a difficult subject. We are dealing with a material whose basic nature is still a matter of discussion. In the case of old glasses, which all contain a wide variety of constituents in major, minor and trace quantities, the situation is far more complex. We have some understanding of the physical and chemical processes which occur when certain glasses, with fairly simple, controlled compositions are subject to carefully designed tests which examine their response to aqueous attack, temperature changes, mechanical stress, etc. Ancient glasses have been exposed for years, sometimes thousands of years, to uncontrolled and often unknown conditions: our limited scientific knowledge is quite unable to interpret many of the results of such an exposure. In short, there are too many variables. Whether we will ever be able to explain, and so perhaps control, the processes of decay in ancient glass is open to doubt: meanwhile we have to deal in an empirical way with the results of that decay.

The philosophy of conservation

If the technical side of conservation is an area of dispute the philosophical side is just as complex. It is obvious that the conservator should respect the integrity of the object, using identifiable techniques that are reversible wherever possible which will allow later treatment if this becomes necessary. However, we are all products of differing cultural environments which affect the way in which we wish to see objects preserved. A good example is provided by our heritage of medieval stained glass, increasingly under attack from the polluted atmosphere, vibrations from heavy traffic, and the sustained weathering effects of rain.

Nowadays the sensitive restorer may maintain a "bank" of pieces of old stained glass which can be used to provide suitable replacements. This is often thought to be the best way to replace missing heads or other obvious features which attract the attention of the observer. Sometimes the area may be filled with plain glass so that there is no possibility of mistaking modern for medieval work. New decorative painting may be introduced but unless carried out with great sensitivity these techniques can produce incongruous and misleading effects. Again, it is common to see restored windows where colours have been used which were never available to the medieval glassmakers. Their range of colouring agents was more limited and their glasses, containing impurities which are not present in modern glasses, look quite different. However, we should not automatically assume that what we see now is what there was at the beginning. These windows have undergone centuries of weathering: even where decay has been slight there is a patina of age which affects the colours that we see. The end result, although most beautiful, may be quite different from the original intention of the artist.

These considerations apply equally to glass which is perhaps of more interest to the archaeologist, namely whole objects or fragments found during the course of excavations. The outstanding characteristic of glass in this case is its brittleness and general fragility, even for pieces which have apparently undergone little deterioration and are of relatively solid construction. This should always be borne in mind by anyone who has to deal with glass from excavations. Although many techniques have been developed for cleaning, consolidation and preservation, no technique has universal applicability and most should be left to experts with long experience in dealing with a wide range of conservation problems.

Treatment of glass finds

Glass objects are found under a wide range of conditions from the extreme dryness of the desert to the saturated surroundings of waterlogged environments. Although very unstable glass will long ago have deteriorated and disappeared that which remains will vary in condition from apparently perfect to material which can be mistaken for rock or pottery and which may only be held together by surrounding soil. Because of their complex compositions and the differing circumstances to which they are exposed glasses can show many types of weathering. These include dulling, fine cracking, frosting, iridescence, crusting, and pitting.

Variations in durability between different types of glass can affect our perception of the period in question, because remaining glass may not be characteristic, in quantity or type, of the original production. For example, the Romans produced a relatively stable soda glass, whilst medieval glass, where the alkali was usually potash, was more susceptible to attack and is therefore scarce in comparison with excavated Roman glass. Often the weathering of soda glass proceeds in layers which form parallel to the surface, of a thickness comparable with the wavelength of light, leading to an iridescent appearance. Layers can easily flake off, leading to thinning and final disintegration of the object, and such specimens must be treated with extreme care especially if they have to be removed from surrounding soil. Weathering of potash glasses often begins at localized centres or at scratch sites and advances on hemispherical fronts giving a characteristic pitted appearance. In severe cases practically none of the original glass structure is retained producing an object of extreme fragility which may have to be consolidated on-site.

Careful recording of the glass before removal is especially important as you can never be certain that it will survive this process in its original condition. For buried objects it may be best to remove them very carefully with plenty of the surrounding soil, bearing in mind that this may be the only thing that is holding them together. There are no completely satisfactory methods of on-site consolidation although various techniques have been used successfully in particular cases. Resins or waxes, which set to form a matrix holding the glass fragments together, have been employed (see, for example, Bimson and Werner, 1972) but great care has to be taken to see that the agents used do not alter the physical and chemical characteristics of the glass.

When glass is removed from its surroundings it can undergo rapid changes in appearance often indicative of a serious alteration in its state. If the glass has been in a wet or damp environment it may rapidly lose its moisture on exposure to sun and air: the equilibrium established over a very long period is destroyed and the glass may even crumble. Glass from very dry surroundings may absorb atmospheric moisture leading to attack and decay. Therefore care should be taken to maintain the relative humidity of the environment during and after excavation and removal.

Very often soil and other surface dirt clings to the glass. It is best to use only a soft brush for removal of this material as scraping can damage the surface and remove weathering crusts. Less fragile samples which have not been excavated under very dry conditions can sometimes be gently wiped with a cloth (linty material should not be used as it can catch on the surface) moistened with distilled water. Methylated spirits or acetone can also be used but all cleaning treatments should be tested on a small sample if possible as even the gentlest method can have unpredictable effects on old glass. Ultrasonic cleaning is probably one of the methods least likely to lead to damage. Newton and Gibson (1974) found that periods of treatment as long as six minutes did not seem to damage well-fired decoration on medieval glass.

After the dirt has been removed the glass remains in its weathered state, and you will have to decide whether to remove the surface layers. There is a temptation to do so because, particularly where "shells" of weathering are present, you may be able to expose an apparently perfect surface. However, by doing so you will alter the dimensions of the object, remove surface decoration and increase the probability of collapse. Crusts flake off all too easily and objects should be kept apart: naturally sharp edges of samples can cause scratching and chipping. For transit a polythene bag of the thin variety may be used for each piece of glass. I have also found that the tissue paper used for dress patterns is useful for this purpose, as it is not too rigid but has a smooth surface which will not catch on the glass. It can also be packed between samples to stop them moving around during transit.

Once the object has been cleaned it needs to be protected from attacking agents which will cause further deterioration. Although water vapour attacks glass rather slowly (at perhaps one hundredth of the rate of liquid water) it can still cause decay so the object must be protected against both condensation and high humidity. Various types of applied surface films have been used, made from organic resins or thin inorganic films but in some cases atmospheric moisture can still penetrate. Such treatments may also drastically alter the nature and appearance of the surface and can themselves degrade with age. Even if the original surface appears to be sound, very fine hair cracks may be present and in these cases application of a coating can result in an accelerated rate of decay. In any case the behaviour of coatings is always unpredictable: Newton (1974) has tested a range of coatings on substrates of modern plate glass and poorly-durable medieval glass and has shown that the coatings on different substrates respond in quite different ways. Thus their use as protective agents over a long period of time is open to question.

Recently experiments have been carried out on the application of thin polymer-like films to simulated medieval glasses by the glow discharge process (Kny and Nauer, 1981). Preliminary results indicate that such films are able to retard the corrosion of these glasses considerably. In the present experiments thin (80 nm) films of tetramethyl tin were deposited on the polished surfaces of model glasses representing average medieval glasses with high vulnerability to environmental attack. The corrosion of coated and uncoated samples was tested in an humidity chamber by exposure to air and by immersion in water. All the uncoated samples showed rapid deterioration after short exposure times, but the coated samples suffered much less damage.

Although this work is at an early stage, there are various features of glow discharge films that seem to hold promise for the protection of poorly-durable glasses. The film composition can be varied within wide limits and thus its thermal expansion could probably be matched with that of the glass. Medieval glasses, with high alkali contents, have a high coefficient of thermal expansion and any mismatch between glass and film may cause a failure in adhesion. The organo-tin films tested by Kny and Nauer had exceptionally good adhesion, probably owing to the formation of tin oxide bonds at the glass/film interface. Glow discharge films also seem to be more successful than resin coatings in preventing moisture from reaching the glass (by diffusion or through cracks in the coating) and subsequently attacking it. As they are usually transparent and applied in very thin layers they do not affect the appearance of the glass, and they are inexpensive and easy to apply.

Such coatings hold promise for the future but further work remains to be done before they can be accepted as reliable protection for unstable glasses. At present it is probably better to store the glass as it is under suitable conditions of humidity. These will depend upon the original condition of the glass. Very high humidity can damage glass over a period of time, especially if the object came from a dry environment. On the other hand some glass which has been in moist surroundings over a long period of time, although suffering a degree of surface deterioration is in a state of metastable equilibrium. If it is then kept in a very dry environment the surface layers dehydrate and the glass quickly deteriorates. A 45–55% range of relative humidity is considered to be reasonable but this may not apply to all glasses for storage and display. Temperature extremes should also be avoided and the glass kept in a cool, dust-free environment.

Bibliography

Bimson, M., and Werner, A. E. (1972). Notes on a suggested technique for the consolidation of fragile excavated glasses. In: *Proceedings of the 9th International Congress on Glass, Versailles, Sept.–Oct. 1971: artistic and historical communications*, pp, 63–92. (International Commission on Glass)

Kny, E., and Nauer, G. (1981). Retardation of the corrosion of medieval glasses by glow discharge films. *Glass Technology* **22**, 38–41.

Newton, R. G. (1974). A problem arising from the weathering of poorly-durable glasses. In: *Proceedings of the 10th International Congress on Glass, Kyoto, July 1974*, part 2, section 9, pp. 9.49–9.54.

Newton, R. G., and Gibson, P. (1974). A study on cleaning painted and enamelled glass in an ultrasonic bath. In: *The deterioration and conservation of painted glass: a critical bibliography and three research papers* (R. G. Newton), pp. 70–78. (Oxford University Press)

Newton, R. G. (1981). A summary of the progress of the Ballidon glass burial experiment. *Glass Technology* **22**, 42–45.

It is difficult to give a range of representative references which will offer practical help to archaeologists because there are so many views on glass conservation. A useful set of review articles which summarizes the various approaches is provided by the book *Conservation in archaeology and the applied arts: preprints of the contributions to the Stockholm Congress, 2–6 June 1975* (International Institute for Conservation of Historic and Artistic Works, 1975). The section on glass contains papers on all aspects of conservation: cleaning, deterioration, restoration, storage, and the mechanisms of deterioration are discussed fully for many types of glass and each paper contains additional references.

As this is not a book for the professional conservator Chapter 5 has been limited to what I hope will be helpful suggestions for archaeologists who have to deal with fragile glass objects. At the time of writing a book on the conservation of glass, by R. G. Newton, is in preparation which will offer comprehensive advice for those who wish to go into the subject more deeply. Newton has also prepared a valuable annotated bibliography *The deterioration and conservation of painted glass: a critical bibliography and three research papers* (Oxford University Press, 1974) which summarizes and comments on the field of glass conservation.

To keep up with the ever-changing glass conservation scene perhaps the most useful source is the *CV News Letter* (Comité Technique du Corpus Vitrearum. 1972–). This publication, at present issued twice-yearly, reports on the progress of glass conservation experiments, news from different groups concerned with conservation, and related topics, as well as containing general articles on glass history. Especially helpful is the abstracts section which collects together references to articles of interest from a wide variety of publications.

There are other, more general publications on conservation which contain information on glass. *Art and archaeology technical abstracts* (formerly *I.I.C. abstracts*) contains informative summaries on the conservation of various types of material, including glass and ceramics. *Studies in conservation* (International Institute for Conservation of Historic and Artistic Works. 1952–) contains occasional articles on glass conservation. *The conservation of antiquities and works of art: treatment, repair, and restoration* 2nd ed., 1971, by H. J. Plenderleith and A. E. A. Werner (Oxford University Press) is intended as a guide and handbook for the archaeologist and curator. There is a section on siliceous materials which contains a chapter on glass. This gives practical advice on deterioration and its treatment, and the repair of broken glass objects. *Recent advances in conservation: contributions to the I.I.C. Rome conference, 1961,* 1963, edited by G. Thomson (Butterworth, for the International Institute for Conservation of Historic and Artistic Works) is a comprehensive survey of areas where positive progress has been made. A section is devoted to the examination and conservation of glass.

For further advice you can also contact The Corning Museum of Glass, Corning, New York, 14830, USA, who can put you in touch with expert conservators.

6

Glass and Glassmaking Sites in Britain

Prehistoric and Roman glass

Although glass finds of Bronze and Iron Age date, particularly of faience beads and objects decorated with glass inlays are widespread in Britain it is uncertain whether these were made locally or were imported. Margaret Guido (1978) has published a systematic and detailed survey of glass beads of the prehistoric and Roman periods in Britain and Ireland, and she concludes that beads were made in Britain, although actual "glassworks" have not been discovered. The conclusion is based on the fact that the distribution of certain bead types is so concentrated in particular localities that they must have been bead-producing centres. One of these sites is at Meare, near Glastonbury, Somerset, where enormous quantities of beads have been found. As Meare is near to the coast it is possible that it was a distribution centre for foreign goods, but because of the distinctive styles of the beads (one speciality was translucent, colourless glass decorated with opaque yellow motifs) it is highly likely that they were of local origin. At nearby Glastonbury beads have been excavated that are very different in character and local production again seems likely. Lumps of fused glass of various colours, as well as beads, have been excavated at Culbin Sands, near Inverness, and Traprain Law, East Lothian: at the latter site a number of small crucibles containing coloured glass have also been discovered. The glass could have been melted from raw materials or fashioned from existing glass pieces, but the variety and complexity of these ancient beads indicate that a considerable degree of skill was involved in their manufacture.

Although much, perhaps most of Roman glass (and later glass) was imported, it seems that everyday glass for glazing and vessels may well have been produced here: glasshouse sites of this period have been discovered in Cheshire, Warwickshire, Norfolk and Salop, but little structural detail is apparent. Whether original raw materials, sand etc. were melted, or whether these were glass working sites, fashioning glass into the desired objects, is open to question.

Anglo-Saxon and Viking glass

Following the Roman withdrawal the glass styles became plainer and a large number of simple vessels of the Anglo-Saxon period have been found, often as part of burial groups. These finds are described in detail by D. B. Harden (1978) in his review article on Anglo-Saxon and later medieval glass. One of the most interesting recent discoveries took place at the Anglo-Saxon monastic houses of Monkwearmouth and Jarrow in Northumbria. Excavations were started by Professor Rosemary Cramp in 1959 and quantities of window glass have since been found as well as mosaic rods and inlay plaques in the extensive excavations that have been carried out. The glass is especially interesting because of the account of it that we have from the Venerable Bede. Bede, a monk at Jarrow who was writing during the first part of the eighth century says that when the monastery at Wearmouth was being built, in 675 AD the founder and first abbot, Benedict Biscop brought over glassmakers from France to glaze the church and to teach the English how to do this for themselves. In 758 AD Cuthbert, the Abbot of Wearmouth, again sent abroad, but this time for glass vessel makers. The two monastic houses were sacked by the Vikings in 867 AD and organized monastic life ceased on these sites for over 200 years. The excavated glass should therefore date to a period between the late seventh and mid-ninth centuries.

The glass is of the soda-lime variety and is generally weathered. It was made by the cylinder process (see Chapter 2) and is of various colours, mainly greens or bluish-greens, also pale blue, olive green, amber, yellow-brown, and red: some of the pieces show streaks and lines. The glass pieces, or quarries, were specially shaped and were probably set in leads in a pattern to form the window. Further details are given by Professor Cramp (1969). No firm evidence for a glass furnace has been found at the sites, although it is almost certain that the window glass was made there. However, at another monastic site, Glastonbury Abbey in Somerset, parts of a glass workshop have been recovered underneath the late medieval cloister and chapterhouse: these may date back to the mid-tenth century when Abbot Dunstan was restoring the abbey. Remains of three furnaces were found, two probably used for annealing and one which might have been used for glassmaking. There were also fragments of small (15 cm high, 18 cm diameter) pots with green glass adhering both inside and outside.

At the present time exciting finds are being made in York where over the past few years archaeologists have been excavating the Viking town and revealing a rich and complex pattern of city life. They have discovered what is believed to be the first evidence of a Viking glass furnace (Addyman, 1981). The bulk of glass-working debris from the Coppergate site consists of potsherds or tile fragments covered or splashed with a glassy material which often has the appearance of a deliberate glaze. A few prices, including one large base sherd, are covered with a glass layer up to 1 cm thick and some of the tile fragments could well have formed part of a glass kiln. Two pieces have been examined by experts at the Corning Museum of Glass, New York: each consists of a potsherd coated with a layer of glass to which a layer of gritty material is adhering. The grains of this material appear to be batch constituents, probably quartz sands, and some of the grains have rounded edges as a result of heating. The glassy layer is a deliberately formed glass, not a result of accidental vitrification. Thus glass appears to have been melted here from raw materials, rather than melted down from old Roman glass as had previously been suspected.

Glassmaking in the Surrey–Sussex Weald

As we move into the post-Conquest period we can tell a great deal more about glassmaking activities from excavations. A major source of information is the extensive work by S. E. Winbolt, G. H. Kenyon, E. S. Wood and others on medieval and later glasshouses in the Surrey–Sussex Weald. Kenyon has given a full account of much of this work in his book (1967). Over 40 sites are known, dating from the early fourteenth to the early seventeenth century, almost all in an area about ten miles square. More than three-quarters of the known glass furnaces are in the old Chiddingfold, Kirdford and Wisborough Green parishes, very close to the Surrey–Sussex border. Wisborough Green was the centre of Elizabethan Wealden glass production. In this forest area fuel was plentiful and local sand from Hambledon may have been used. Early production was rather crude but during the sixteenth century living standards were rising and there was an increasing demand for good quality vessel and window glass. In the second half of the sixteenth century, under the influence of Jean Carré (see Chapter 2) the industry increased in size by four or five times. Kenyon points out that the woodlands of only three parishes supported the early industry for over 300 years, thus indicating its small size.

After 1567, and until the industry moved away over 50 years later there were increasing quarrels and petitions to the authorities over disputed rights. Wood for fuel became very scarce, the more efficient coal-fired furnace was developed and during the period 1615–8 the conflict between the wood-fired Wealden forest glasshouses and the coal-fired town glasshouses became more acute. In 1615 an official edict had been proclaimed prohibiting the use of wood for firing glass furnaces and insisting on coal instead: it was Robert Mansell's determination to enforce this edict that finally drove away the Wealden glassmakers.

Glasshouses at Blunden's Wood, Knightons and Bagot's Park

It is impossible to know how many glasshouses there originally were, owing to later extensive ploughing and planting: Kenyon says that the furnaces were little larger than bread ovens, with a working space roughly roofed over, with timber supports, and showing no standard design. Only a very few so far have yielded useful structural evidence, and of these one of the most important is the earliest known, Blunden's Wood, near Hambledon, Surrey. This was completely excavated by E. S. Wood in 1960 and the work is described by him in a later paper (Wood, 1965).

There were three furnaces at the Blunden's Wood site. Figure 11 shows a plan view of the site with the main furnace (A) which consisted of a straight central flue with firehearths at each end and stone banks on either side, each with two sieges, or benches, where the glass melting pots would have been placed in slight depressions. The furnace was eleven feet long and had slightly bowed walls, being ten feet wide at the west end, eleven feet at the centre and eight feet at the east end. No coherent evidence of the roof remained but it may have been barrel shaped. The flue was blocked with grey material which turned out to be solidified glass: this would have overflowed from the crucibles and finally blocked the flue, stopping the necessary draught and leading to the abandonment of the furnace. Figure 12 is a sectional view of a suggested reconstruction of the furnace (Ashurst and Wood, 1973).

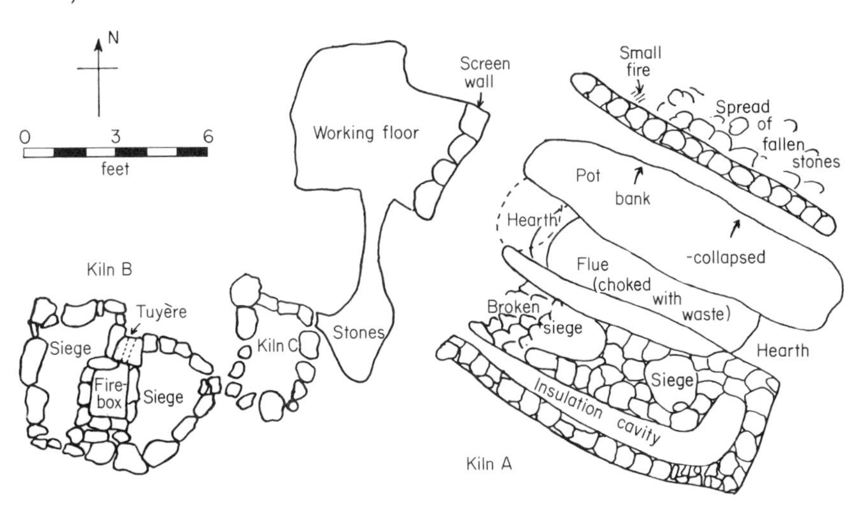

Fig. 11. Blunden's Wood site: general excavation plan. Reproduced by courtesy of Mr E. S. Wood.

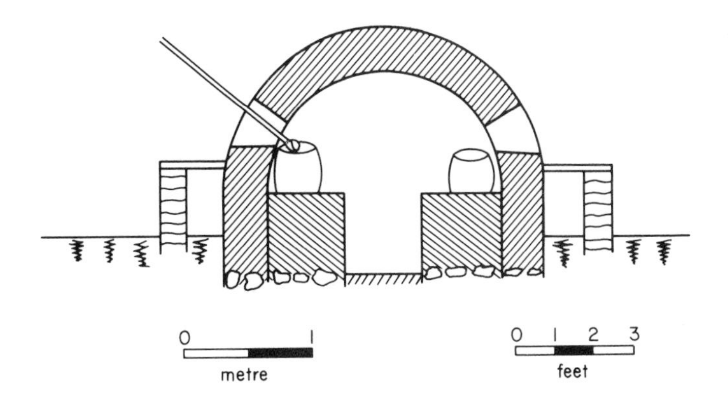

Fig. 12. Blunden's Wood main furnace: sectional view of suggested reconstruction. Reproduced by courtesy of Mr E. S. Wood.

Seven feet to the west of the large furnace was another, roundish furnace (B) containing a fireplace and between the two furnaces was a paved area forming a working floor, and the remains of a third small round oven (C). Wood considers that this oven may have been used for preheating pots, the other small furnace for fritting and annealing (all of which would be carried out at lower temperatures than those required for melting) and the main furnace for melting. Glass found was pale green in colour and formed fragments of windows and vessel glass. Potash was the predominant alkali and the lime content was high, consistent with manufacture from local sand and wood ashes: samples showed considerable weathering, typical of such a composition. Tests were also performed on fragments of melting pots and it was found that the clay would withstand temperatures well above the glass founding temperature (1200–1250°C). The clay was not local and this is rather strange as Blunden's Wood clay is perfectly capable of withstanding such temperatures. Wood speculates that the pottery and glass industries were clearly demarcated and that one regularly supplied the other.

The method of dating the Blunden's Wood furnace is interesting. Pottery fragments found at the site assigned it to the first half of the fourteenth century, but the method of thermo-remanent magnetic dating (also known as archaeo-magnetic dating) was applied to samples from the main furnace. The method relies on the fact that the earth's magnetic field is constantly changing, both in direction and magnitude. When clay (which contains the magnetic material iron) is subject to high temperatures and then cooled it retains the magnetic field that existed at the time of the last firing. If the fired clay remains undisturbed it is often possible to measure this and to compare it with values for other sites of comparable age where the date is known. Curves can be constructed plotting certain values associated with the magnetic field for sites of known age and the age of the unknown site read off. The values for Blunden's Wood lay midway between those of two structures of comparable age (a pottery and a tile kiln) dated by other methods to 1307 and 1350. This gave a reliable date of c. 1330, in agreement with the evidence of the pottery. Blunden's Wood is typical of the forest glasshouse of "northern" type (see Chapter 2), a layout that seems to have altered little before the late sixteenth century. The building was rectangular and the melting pots stood on raised banks on each side of a flue.

Furnaces with similar plans, dating from the early 1550s, have been excavated at Knightons, Alfold, Surrey. A plan of the site is shown in Fig. 13. Three main furnaces, all built on a rectangular plan, have been excavated and the second furnace overlaps the earliest by some two or three feet. The third furnace probably replaced the second after only a few years. Connected to the third furnace is a two-chamber annealing furnace containing large quantities of crown glass (originally stacked up in sheets nearly three feet across): the glass is of good quality and is a very early example of crown glass manufacture in

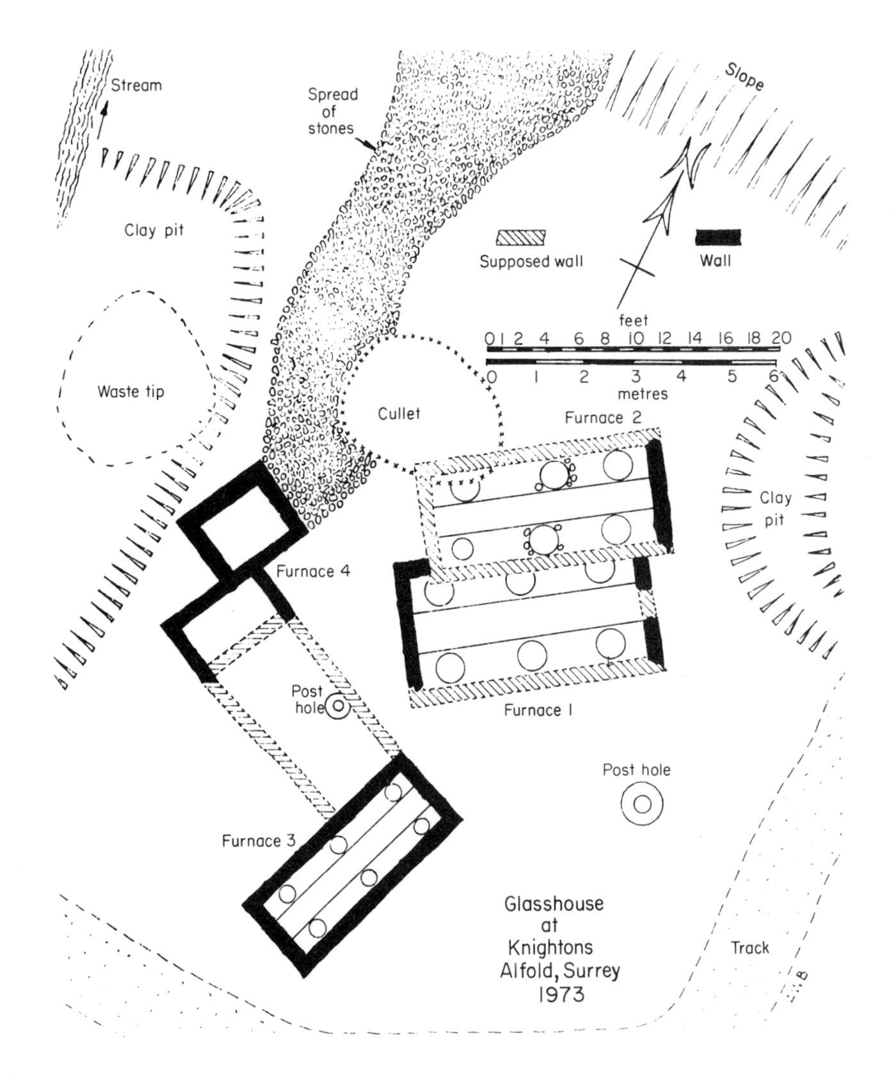

Fig. 13. Knightons, Alfold, Surrey: general excavation plan of site. Reproduced by courtesy of Mr E. S. Wood.

England. As well as window glass Knightons produced flasks and specialized glassware such as alembics, urinals and hour-glasses.

Other examples of the rectangular glasshouse are known in Surrey and Staffordshire. In particular 15 glassmaking sites have been found at Bagot's Park to the north of Abbots Bromley, Staffordshire during the course of land reclamation: one of these has been excavated by David Crossley (Crossley, 1967). The two furnaces found on the site have been firmly dated to the beginning of the sixteenth century, but it is believed that at this time there was already a tradition of glass-making going back for several generations.

The larger of the two Bagot's Park furnaces was built of brick and stone with a clay dome and was used for melting glass in six crucibles placed on sieges on each side of the flue, with stoke holes at opposite ends. At each end of the furnace was a pair of holes capable of taking posts fifteen inches in diameter, suggesting an all-over roof covered in tiles (remains of which were also found). The whole structure may well have looked very like the fifteenth century glasshouse shown in Fig. 7, Chapter 2: Bagot's Park was the first site to show evidence for such a tiled roof over the main furnace. A second, smaller furnace nearby had not been subject to very high temperatures and was probably used for annealing the finished objects.

Hutton and Rosedale glassmaking sites

Towards the end of the sixteenth century a new shape of glasshouse began to appear. Examples are known from various parts of the country but some of the best ones come from North Yorkshire and their excavation is described in the paper by Crossley and Aberg (1972). The furnaces, at Hutton and Rosedale in the (then) North Riding were studied during the period 1968 to 1971 and are of the type which can be described as "wing" furnaces because of the wings attached to the corners of the main central furnace. A plan of the Rosedale furnace is shown in Fig. 14 and the difference between this and the rectangular northern forest glasshouse is apparent.

Both sites, in valleys running southwards from the North Yorkshire Moors, were well chosen from the point of view of raw materials: timber, sand, fireclays and stone were easily accessible. The Hutton Common site, about one mile south of Hutton-le-Hole, is in an area where deciduous trees were probably plentiful in the sixteenth century, and at Rosedale the valley bottom is still well wooded. Initial surveys were carried out with a proton magnetometer: this sensitive instrument registers and measures magnetic anomalies, slight variations in the earth's magnetic field. Structures which have been subject to high temperatures produce such anomalies and thus the method is very useful for locating remains of buried glass furnaces.

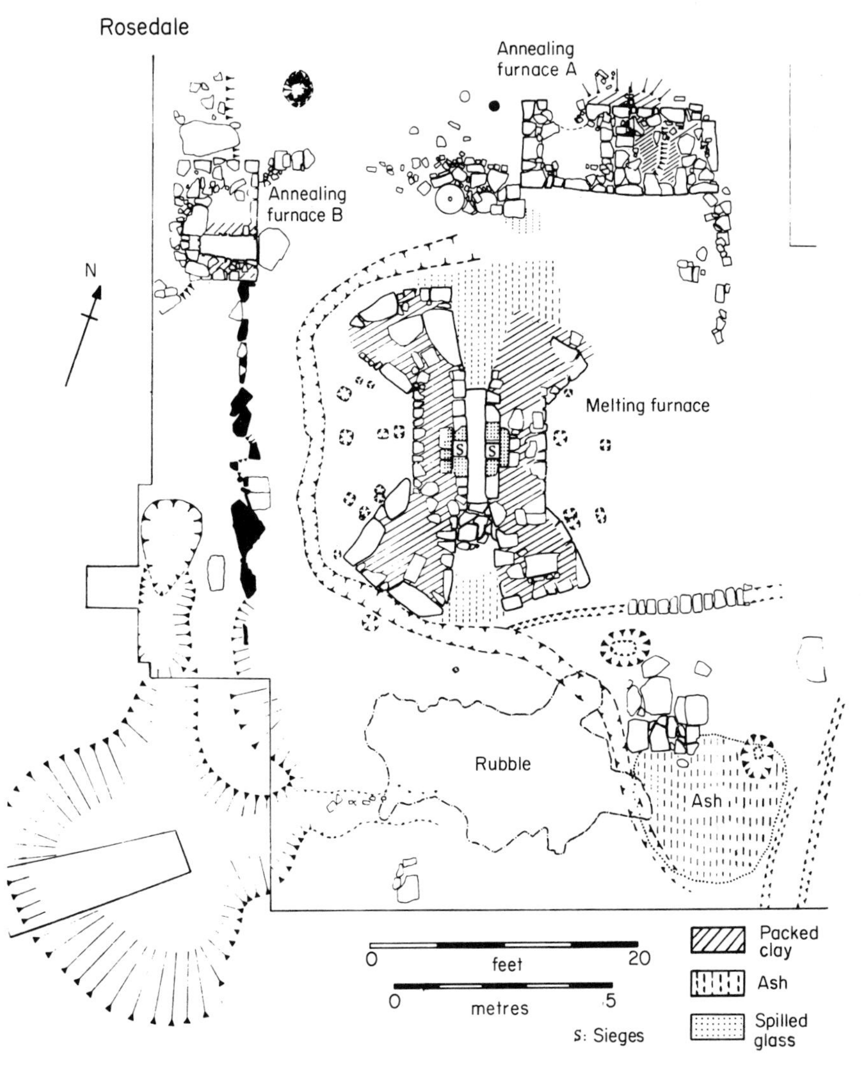

Fig. 14. Rosedale: plan of glass furnaces. Reproduced by courtesy of Mr D. Crossley.

A plan of the main Hutton furnace is shown in Fig. 15. The excavations on the furnace showed that there were three main periods of construction and use. The earliest structure was rectangular with two siege blocks, each probably carrying a single crucible, but most of the structure was dismantled and the furnace rebuilt during the second phase when the wings were added to the south west and north east corners. During the third period the central furnace was reconstructed but it retained the same general outline with a pair of stone sieges each having a space for one crucible. The remains of a small furnace were found to the west of the main furnace which may have been used at an early stage for annealing the glass.

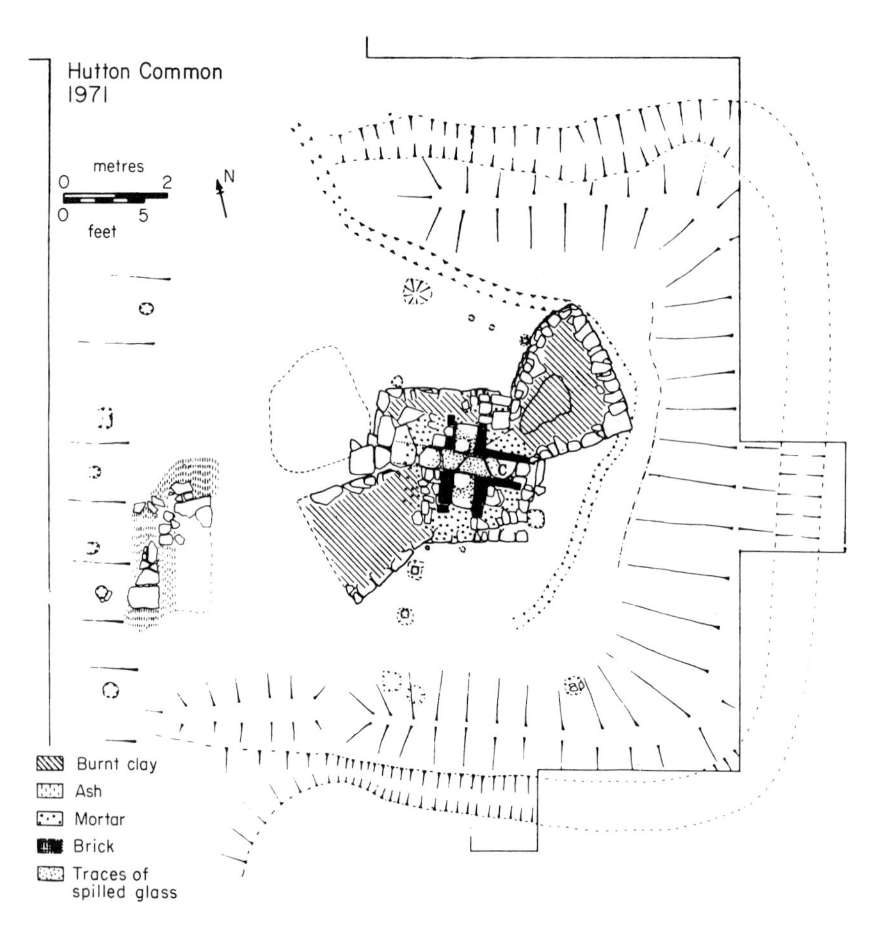

Fig. 15. Hutton Common: plan of site showing main furnace. Reproduced by courtesy of Mr D. Crossley.

At Rosedale the whole of the main furnace (Fig. 14) had been built at one period. Sieges on either side of the central flue could each accommodate one crucible and the positions of the stoke holes at the northern and southern ends of the flue were marked by ash and signs of burning. The furnace had four corner wings and, unlike Hutton where the two wings were added at a later stage, they formed an integral part of the original structure. Postholes suggested roofs for the working areas, and remains of two rectangular structures were found nearby which may have been annealing furnaces.

Magnetic dating of the sites suggests that the furnaces were built in the last quarter of the sixteenth century and this conclusion is supported by finds of pottery, coins and a certain amount of documentary evidence. They thus date from the time when Carré was establishing his immigrant glassmakers in the Weald (see Chapter 2) and Crossley and Aberg consider that the winged furnaces are typical of immigrant practice (Wealden sites of the same period also display structural changes from the old forest glasshouse: at Van Copse, Hambledon, a furnace has been excavated having four wings attached to the main furnace). The wings were not used for actual melting but for subsidiary processes such as fritting, reheating whilst glass was being worked, preheating of pots, and annealing the finished articles. They operated at a lower temperature than the main furnace: this was arranged by having separate fires in the wings or by drawing heat from the main furnace. Hard baked red clay at Hutton may suggest the former arrangement, contrasting with the lighter burning at Rosedale.

Crossley and Aberg mention the account of Christopher Merrett (1662) translating and commencing on Neri's description of glassmaking in his *L'Arte Vetraria* of 1612. It is worthwhile taking this a little further because the arrangements and processes described, although dating from a somewhat later period, reflect a continuing tradition and cast light upon the practices of Hutton and Rosedale.

Merrett describes furnaces of various types with different arrangements for carrying out the melting and subsidiary processes. He is, in fact, commenting not only upon the southern European practices familiar to Neri but the types of glass furnaces described by the German scholar Georgius Agricola in his great work *De Re Metallica* published in 1556. Merrett thus compares the practices of the previous century with those of his own day and it is possible to see how the extensions to the central furnace may have developed.

The basic type of furnace described by Agricola was the "southern" furnace, of beehive shape, described in detail in Chapter 2. Often this furnace was divided by horizontal partitions into compartments, the lower for the fire, the middle for the melting pots and the upper (which was cooler) for annealing the finished articles. Heat passed upwards through holes at the centres of the partitions, thus warming the melting and annealing compartments in turn.

One of Agricola's drawings, however, shows a beehive (two-storey) furnace built against a second, rectangular furnace with a barrel vault roof. Figure 16, taken from the first Latin edition, and first illus-

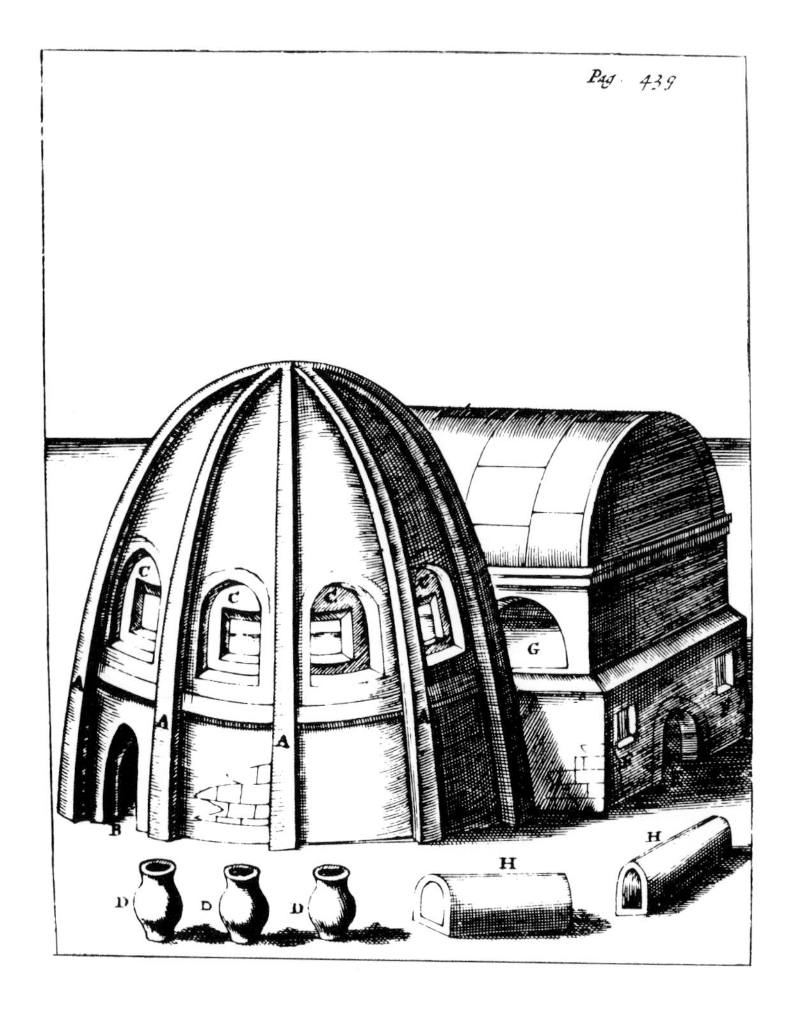

Fig. 16. Beehive and rectangular furnaces for melting and annealing glass. Their relationship is discussed in the text.

trated version of Neri's book (1668) shows the arrangement that Agricola describes. It is not clear from the drawing how the two furnaces were related to one another, but Agricola says that:

> In the back part of the (beehive) furnace is a rectangular hole, measuring in height and width a palm, through which the heat penetrates into the (top half of the rectangular) furnace which adjoins it.

There was also provision for a fire to be lighted on the lower floor of the rectangular furnace, presumably to adjust the furnace temperature if required. The finished articles were placed in earthenware receptacles in the top of the rectangular furnace: "so that they may cool to a milder temperature; if they were not cooled they would burst asunder." Here are the elements of the Hutton and Rosedale furnaces. Further developments are described by Merrett:

> The leer . . . to anneal and cool the vessels . . . comprehends two parts, the tower and the leer. The tower is that part which lies directly above the melting furnace with a partition betwixt them . . . in the midst whereof. . . there's a round hole . . . through which the flame and heat passeth into the tower . . . on the floor or bottom of this tower the vessels fashioned by the masters are set to anneal, it hath two . . . mouths, one opposite to the other, to put the glasses in as soon as made . . . and after some time these glasses are put into iron pans . . . which by degrees are drawn . . . all along the Leer, which is five or six yards long, that the glasses may cool *gradatim*, for when they are drawn to the end of the leer they become cold. This leer is continued to the tower, and arched all along about four foot wide and high within.

It is clear from the text that this long annealing leer was an extension from one mouth of a beehive furnace for melting fine quality glass to be made into vessels. The beehive furnace was not the type in use at Rosedale or Hutton but Merrett also mentions:

> green glass furnaces which are made square . . . having at each angle an arch to anneal their glasses . . . they make fires in the arches, to anneal their vessels.

He also says that the glassmakers worked their glass on two opposite sides of the furnace, and on the two other sides

> they have their calcars (small ovens) into which linnet holes (from the French, lunette) are made for the fire to come from the furnace to bake and prepare their frit, and also for the discharge of smoke . . . so that they make all their processes in one furnace onely.

Merrett does not give details of the size and shape of the annealing sections or the calcars, but it can be seen that Rosedale and Hutton incorporate elements of the structures that he describes. Their central furnaces are typical of those found in European forest glasshouses but new influences are apparent in the wing construction. These developments continued during the seventeenth century and the results were described by Neri and Merrett. The existence of these northern furnaces is evidence for the rapid spread of glassmaking during the sixteenth century, at a time when growing prosperity created a demand for glass products. This made it worthwhile for the glassmakers to move to regions where previously there had been no industry, and the goblets, beakers and bottles produced at Rosedale and Hutton would be sold to a large section of the community. York was at this time a flourishing trading centre and the market for glass products would be further increased by the prosperous farmers of the region.

The coal-fired furnace: excavations at Haughton Green

Rosedale and Hutton were furnaces fuelled with wood which was also a source of alkali in the form of ash: remains of birch, alder and oak were amongst the samples analysed.

The period of wood-fired furnaces was drawing to a close, however, and the seventeenth century saw the development of the coal-fired furnace. The excavations so far of furnaces of this transitional period have not revealed as much information as has work on the earlier forest glasshouses, although it is clear from Merrett's comments that the coal-fired furnaces which he describes must have developed from the structures of the earlier wood-burning furnaces. One of the most interesting excavations of a coal-fired furnace of this period is that at Haughton Green, Denton, near Manchester, excavated by Ruth Hurst Vose and her colleagues between 1969 and 1972 (Vose, 1980). As Mrs Hurst Vose points out, a coal-fired furnace, although producing much higher temperatures (and therefore better quality glass) requires a very good draught to allow the coal to burn properly. Although the complete structural layout is not too clear it is believed that the furnace held four pots and a prominent flue system running beneath two parallel sieges was discovered. Heavily burnt furnace debris showed that high temperatures were certainly attained. We also have contemporary evidence for high temperatures in Merrett's account:

> The stones wherewith the inside of these furnaces are not brick (for these would soon melt down into glass . . .) but hard and sandy . . . The heat of those furnaces, is the greatest that ever I felt, and I have observed straws put in three days after the extinction of the fire soon converted into a flame.

The English cone glasshouse: excavations at Gawber

Coal furnaces spread throughout the country during the late seventeenth and early eighteenth centuries and towards the end of this period the tall cones of the "English" glasshouse became a familiar sight: this was the next stage in attaining even higher temperatures. Progress was uneven in this field because not all furnaces used coal as fuel. As late as the middle of the eighteenth century both plate and sheet glass were produced in wood-fired furnaces and in the absence of legislation prohibiting the use of timber as a fuel wood-firing continued in France for much longer than in England. Although very advanced in other respects, the Manufacture Royale des Glaces, established at St. Gobain in 1693 for the production of plate glass was still using wood in 1819, although the French glassmakers had tried, without success, to substitute coal for wood since 1763. By 1829 they had succeeded in melting in a coal furnace but still had to fine, or remove gas bubbles, in a second, wood-fired furnace. Thus the coal-fired English cone furnace may have been an object of admiration and is given a place in the glass section of the great eighteenth century encyclopaedia of Denis Diderot and Jean Le Rond d'Alembert. Figure 17 shows a plan view of the "Verrerie Angloise" taken from this work: *a* is the main furnace, *b* the grating for the coal, *c* the (four) pots in the furnace, *d* the pot arches used for firing pots before placing them in the heat of the oven, and *f* the small annealing furnaces. The passages between the central furnace and the pot arches for the flow of heat can be clearly seen. Figure 18 shows the outer view of one of these cone furnaces. The tall cone (rising sometimes to over 90 feet and having a basal diameter of 40 feet or more) acted as a tall chimney and ensured an excellent draught, resulting in much higher melting temperatures. Figure 19 shows the work going on inside such a cone: the operations of gathering, blowing, marvering and working at the chair can clearly be

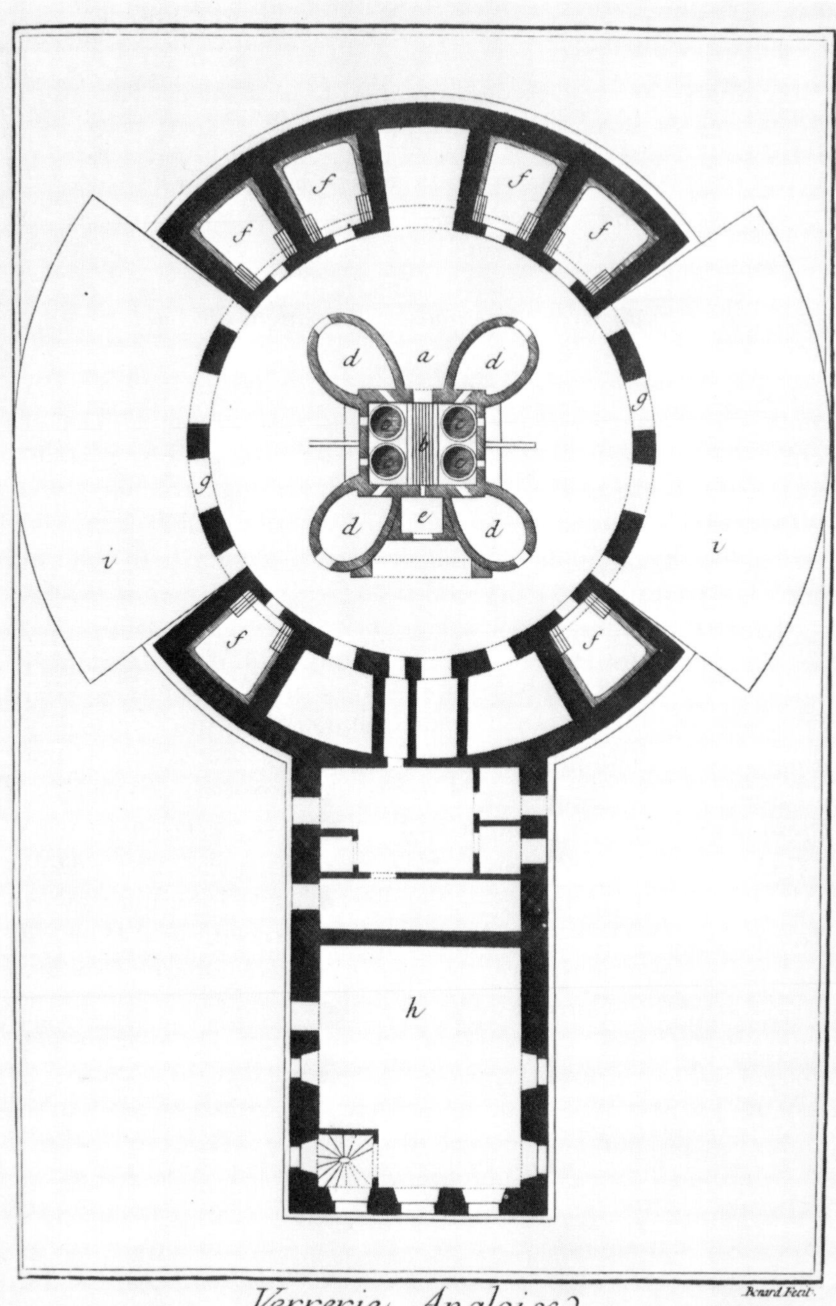

Verrerie Angloise.

Plan du premier étage d'une Halle avec son Four et le Batiment de Service

seen. In this case the pot arches (one of which can be seen at the back of the picture) are not attached to the central furnace. The melting furnace was fired from a fireplace in its centre, below ground level, and air was supplied to the fire via an underground tunnel. The flames rose into the furnace and passed over the pots: combustion products escaped through flues in the side wall and so up through the cone to the outside air. The working conditions were also greatly improved and may be contrasted with the conditions during the seventeenth century:

> They receive scorching heats sallying directly into their faces, mouths and lungs . . . they are for the most part pale, thirsty and not very long lived, by reason of . . . the diseases of their head and breast, and that having their bodies weak, they are soon fudled with wine or beer.

The cone furnaces were a common sight even as late as the early twentieth century. There are now no more than a handful of these beautiful structures left, one of the best preserved being at Lemington-on-Tyne, in a region which was once a major centre of glassmaking. Another, not in such good condition, can be seen at Catcliffe, near Sheffield.

South Yorkshire was a very important glassmaking area in the eighteenth and nineteenth centuries and it is here that major excavations have been carried out by Denis Ashurst and his co-workers. One of his most interesting finds has been at Gawber, near Barnsley (Ashurst, 1970). Although there is documentary evidence for glassmaking in the Barnsley area as far back as the late seventeenth century the origins of the Gawber glassworks are obscure. Thermo-remanent magnetic dating indicated that the earliest excavated furnace was last fired between 1690 and 1735, but glass finds associated with the furnace suggested that it was not operational before the first decade of the eighteenth century. Following this early activity the furnace was not used, the structure was robbed and the core was exposed to weathering prior to levelling and later reconstruction. It was therefore very difficult to reach conclusions about the type of furnace that was built during this early phase but it was probably of the traditional forest glasshouse type. Remains of what appeared to be siege banks, upon which the crucibles would have rested, were found. An interesting

---

Fig. 17. Plan view of an English cone furnace: (*a*) main furnace; (*b*) grating for coal; (*c*) pots in the furnace; (*d*) pot arches for firing pots before placing them in the furnace; (*e*) fritting furnace; (*f*) small annealing furnaces.

Fig. 18. The English cone furnace, familiar feature of the eighteenth and nineteenth centuries. The tall cone ensured a good draught to the coal-fired furnace resulting in higher melting temperatures than were possible with wood-firing. The transport of raw materials, coal and finished products was made much easier by the construction of a widespread network of canals, one of which is shown in the foreground of the picture. Reproduced by courtesy of the Department of Ceramics, Glasses and Polymer, University of Sheffield.

Fig. 19. The interior of an English cone furnace. The glass cones were spacious and provided good working conditions for the workers. Around the walls of the cone can be seen the mouths of the annealing ovens and the "pot arches", or ovens for pre-heating the pots before subjecting them to the intense heat of the main furnace. Reproduced by courtesy of the Department of Ceramics, Glasses and Polymers, University of Sheffield.

feature was the existence of drainage channels between and at the side of the sieges: any water on the site could thus drain away and not remain to form steam with the consequent risk of explosion. This feature is not common, in fact Ashurst and Wood (1973) note that only four forest glasshouse sites have any drains at all, and these form a large rectangle surrounding the entire site. Their function was presumably to carry away water from the roof of the glasshouse and thus

the early furnace at Gawber may represent a distinctive seventeenth century innovation in glassmaking.

Records show that the numbers of glassmakers in the area suddenly increased during the 1730s: this growth continued and by 1818 there were 22 different families working in the industry. It was during this period of activity that the building of the second furnace took place.

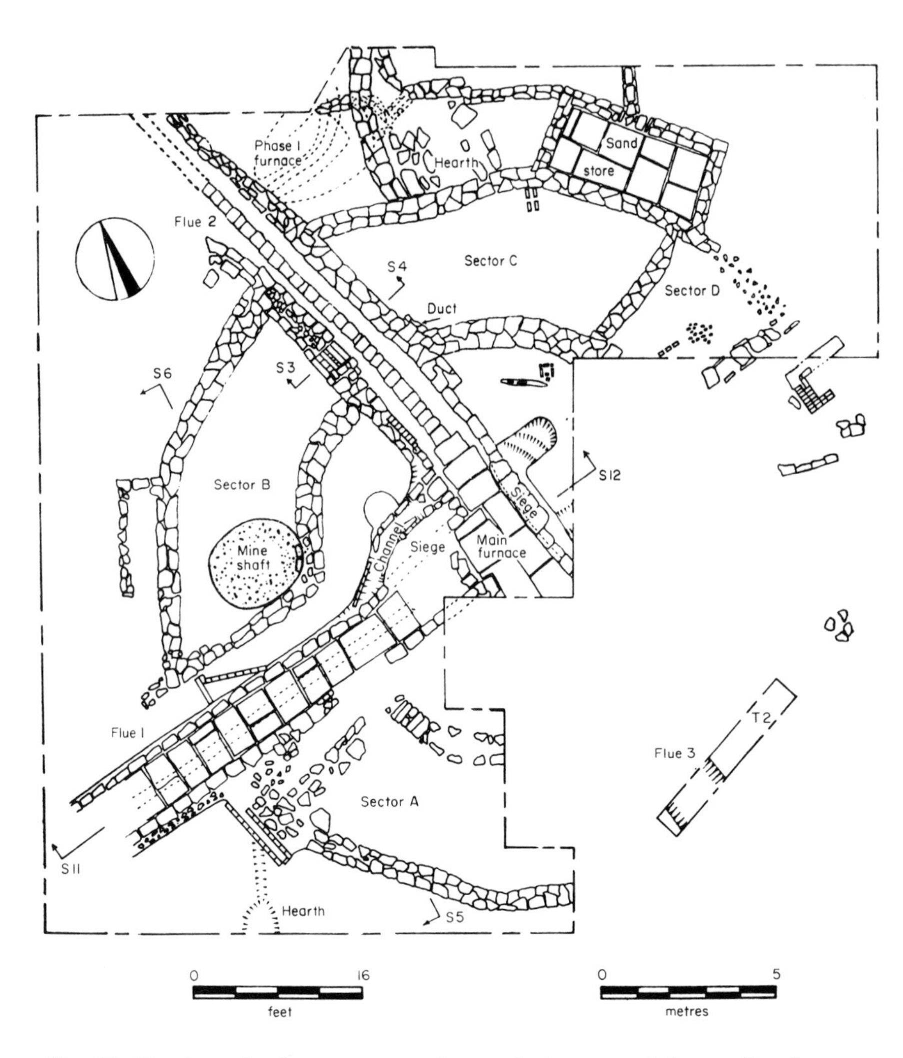

Fig. 20. Gawber glasshouse: cone of second phase overlying earlier furnace. Reproduced by courtesy of Mr D. Ashurst.

Figure 20 shows a plan of the site, with the later phase furnace overlaying the earlier furnace (marked phase 1). Although only foundations remained it can be assumed that the later furnace had a conical brick superstructure similar to the one still remaining at Catcliffe. The plan clearly shows the position of the outer wall of the cone and, ten feet within this, the concentric foundation of what was probably a dome covering hearth and sieges. Three radial flues fed air into the furnace and drains ran down their centre lines. Little remained of the central main furnace, although fire bars were found which fitted into sandstone blocks to form a detachable fire grate.

It is apparent from the differences between the first and second stage furnaces that Gawber represents a transitional phase in glassmaking. Ashurst points out that at this time the traditional forest glasshouse was being replaced by new structures requiring considerable capital investment and capable of producing large quantities of glassware to supply an expanding market. In South Yorkshire the raw materials, coal, clay, stone and sand were freely available, the population of this industrial region was growing rapidly, and so the Gawber glasshouse was well placed to satisfy an increasing demand for bottles, phials and window glass.

In 1821 the Gawber works were sold and glassmaking ceased at the site. Ashurst gives several reasons for this sudden shutdown: these include troubles with excise duties on glass, lack of suitable transport facilities and increasing competition from the growing number of newer works in South Yorkshire. As the nineteenth century progressed new developments in furnace technology were to affect these works, and others like them throughout the country. In the last part of this Chapter I shall describe briefly the changes that took place which led directly to the efficient furnaces of today and the abandonment of the older types of furnace.

The development of modern furnaces

All the furnaces described so far were direct-fired, that is, the fuel (coal or wood) was burnt and the combustion products passed directly over the pots. Although some improvements were made in direct-firing during the eighteenth century the step which led to the modern furnace was the development of the regenerative principle. This in turn depended upon a proper understanding of the nature of heat. Earlier ideas suggested that heat was an invisible, weightless fluid, termed caloric, which material bodies could absorb, their temperatures thereby being raised. However the experiments of Count Rumford and Humphry Davy at the end of the eighteenth century showed that the concept of heat as a fluid, present in a fixed quantity in a particular body at a given temperature, must be abandoned because an unlimited supply of heat could be obtained by rubbing two bodies together for a sufficient length of time. Joule, in his experiments beginning in 1840, then established that a constant amount of heat was produced by a given amount of work. These measurements were the basis for the First Law of Thermodynamics, a statement of the observation that when work is transformed into heat, or heat into work, the quantity of work is equivalent to the quantity of heat. This concept of equivalence led very quickly to improvements in furnaces, brought about largely by the work of the Siemens brothers. Charles William Siemens in particular was influenced by the work of Joule. In 1857 he wrote:

> Our knowledge of the nature of heat has been greatly advanced of late years by the investigations of Mr. J. P. Joule of Manchester, and others; which have enabled us to appreciate correctly the theoretical equivalent of mechanical effect or power for a given expenditure of heat . . . If we investigate the operations of melting and heating metals, and indeed any operation where intense heat is required, we find that a . . . proportion of heat is lost, amounting in some cases to more than 90 per cent of the total heat produced.
>
> Impressed by these views the writer has for many years devoted much attention to carrying out some conceptions of his own for obtaining the proper equivalent effect from heat . . . The regenerative principle appears to be of very great importance and capable of almost universal application.

The regenerative principle had previously been applied by Siemens to various industrial processes, but it is perhaps most clearly described in a patent granted in 1856 to his brother, F. Siemens for the invention of an "Improved arrangement of furnaces, which improvements are applicable in all cases where great heat is required":

> My improvement consists in so arranging . . . furnaces . . . that the products of combustion on their passage from the place of combustion to the stack or chimney shall pass over an extended surface of brick, metal, or other suitable material, imparting heat thereto, which heat serves to heat the atmospheric air or other materials of combustion. The result of this arrangement is, that (they) are nearly heated to the degree

> of temperature of the fire itself, in consequence whereof an almost unlimited accumulation of heat or intensity may be obtained.

In the furnace four chambers were built containing brickwork with channels through which gases could pass. Through one pair of these chambers the combustion products were passed to their final exit through the chimney: these hot gases heated the brickwork and after about half an hour a reversal valve caused the combustion products to pass through the second pair of chambers while the incoming gas and air each passed through one of the first pair of chambers. The cycles continued, the heat of combustion was said to be regenerated and the chambers were known as regenerators. In the original design the fireplaces for gasification of the fuel (by a process of partial combustion and volatilization of solid fuel) were built into the furnace, and so smoke and dust were taken through the regenerators with consequent deleterious effects on the glass.

Steps were quickly taken by the Siemens brothers to correct this fault and in 1861 a further parent was granted to C. W. and F. Siemens for "Improvements in Furnaces". Essentially it incorporated a unit for combustible gas production separate from the main body of the furnace:

> . . . so that the introduction of solid fuel into the glass furnace may be altogether avoided, and the gaseous fuel may be heated to a high degree prior to its entering into combustion with atmospheric air, also heated to a high degree, thus causing great economy of fuel. There is also great advantage derived from the absence of any solid carbon or ashes in the working chamber of the furnace . . . We are thus enabled to melt flint, extra white and other superior qualities of glass in open pots . . . without injury.

(It should be noted that covered pots, to protect the glass, had been introduced at the end of the eighteenth century for the production of fine plate glass.) The Siemens regenerative furnace, applied to glass manufacture, is shown in Fig. 21.

Despite problems such as soot accumulating in the flues and dropping from the roof of the furnace, which was caused by the intense heat, the trials of the 1860s, carried out largely at the works of Chance Brothers, Birmingham, were extremely successful. It was possible to use much higher temperatures and save a great deal of fuel. The process was made even more efficient by the substitution of pots for melting the glass by a glass "tank". The reason for this is that in a pot furnace the melting chamber is not used efficiently. Fuel is burnt outside the pots and heat is transferred to the melt through the pot walls. Experiments to get rid of pots seem to have started in the 1840s but it was not until 1860–1 that Frederick Siemens built his first tank furnace. He and his brother Hans (living in Germany) carried out a great deal of development work and in 1867 Frederick succeeded in converting the tank from intermittent operation (melting batch at night and working during the day) to continuous operation. The continuous tank consisted of three parts. Batch was charged into the melting chamber and was melted by gas flames sweeping across the surface. The molten glass then sank to the bottom of the chamber and passed through verti-

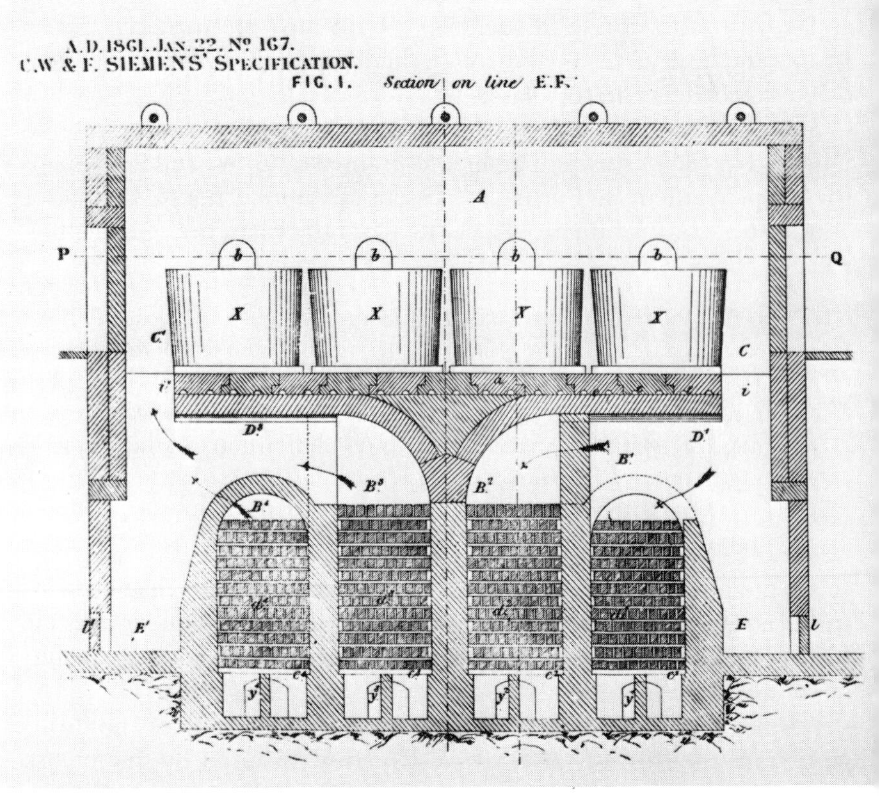

Fig. 21. The Siemens regenerative furnace, as applied to the manufacture of glass. A longitudinal section is shown. B^1, B^2, B^3, B^4, regenerators filled with checker-work of fireclay bricks; *a*. furnace floor; *x*. pots; *b*. ports in furnace for working pots; C^1, C^2, C^3, C^4 gratings connecting regenerators to apertures at their bases for the passage of air and gas.

cal channels into the top of the second chamber, where refining (removal of small gas bubbles) took place. After refining the glass moved into the third chamber, the working end of the tank.

The introduction of the continuous tank furnace was a radical departure for glassmaking. The old multi-stage, intermittent melting process, unchanged in essentials for many hundreds of years, was now superseded. With a continuous regenerative tank furnace regular operation was possible with an output per shift of about 75–80 cwt of glass: the production capacity was about double that of the ordinary pot furnace. The continuously maintained melting temperature saved fuel because no heat was lost during a period of cooling: also the wage bill was reduced by about 60% as skilled workers for batch charging were no longer required.

The Siemens furnace, unchanged in essentials, is used in modern glassmaking, and so if you are interested in developments in glass manufacture since the abandonment of the eighteenth century glasshouse you can still learn a great deal from watching a modern furnace in operation. A more detailed account of the technical developments in glassmaking during the nineteenth and twentieth centuries can be found in the book by R. W. Douglas and S. Frank (1972). To complement this and to learn about furnaces from earliest times up to the eighteenth century you can consult the article by R. J. Charleston (1978): this comprehensive account is copiously illustrated with original drawings, engravings and plans.

The adoption of the regenerative tank furnace was rapid, especially in the glass container industry: in 1872, out of 177 furnaces in the country, 40 were tank furnaces. The workers did not accept this new development readily because it represented a considerable threat to their livelihoods. Nevertheless, when mechanical methods of bottlemaking were introduced, and sheet and plate glass began to be made by continuous processes, the regenerative tank furnace became an essential part of the complete plant and still remains at the heart of modern glass manufacture.

Bibliography

Addyman, P. V. (1981). Private communication on excavations at the Coppergate site, York.

Ashurst, D. (1970). Excavations at Gawber glasshouse, near Barnsley, Yorkshire. *Post-Medieval Archaeology* **4**, 92–140.

Ashurst, D., and Wood, E. S. (1973). Glasshouses at Gawber and Blunden's Wood: a further note. *Post-Medieval Archaeology* **7**, 92–94.

Charleston, R. J. (1978). Glass furnaces throughout the ages. *Journal of Glass Studies* **20**, 9–33.

Cramp, R. (1969). Glass finds from the Anglo-Saxon monastery of Monkwearmouth and Jarrow. In: *Studies in glass history and design: papers read to Committee B sessions of the 8th International Congress on Glass held in London 1st–6th July, 1968* (R. J. Charleston *et al.*, eds), pp. 16–19.

Crossley, D. W. (1967). Glassmaking in Bagot's Park, Staffordshire, in the sixteenth century. *Post-Medieval Archaeology* **1**, 44–81.

Crossley, D. W., and Aberg, F. A. (1972). Sixteenth-century glass-making in Yorkshire: excavations at furnaces at Hutton and Rosedale, North Riding, 1968–1971. *Post-Medieval Archaeology* **6**, 107–157.

Douglas, R. W., and Frank, S. (1972). *A history of glassmaking.* (Foulis)

Guido, M. (1978). *The glass beads of the prehistoric and Roman periods in Britain and Ireland.* (The Society of Antiquaries of London)

Harden, D. B. (1978). Anglo-Saxon and later medieval glass in Britain: some recent developments. *Medieval Archaeology* **22**, 1–24.

Kenyon, G. H. (1967). *The glass industry of the Weald* (Leicester University Press).

Vose, R. H. (1980). *Glass.* (Collins), pp. 143–146.

Wood, E. S. (1965). A medieval glasshouse at Blunden's Wood, Hambledon, Surrey. *Surrey Archaeological Collections* **62**, 54–79.

7

Glass and its Place in Archaeology

Introduction

A glass object often has an intrinsic attraction, or it may simply be an interesting sample, but its importance to the archaeologist lies in what it can reveal about the past. This depends partly on the context in which it is found and partly on the specimen itself. The importance of context is stressed in archaeological textbooks, so I shall concentrate on the second aspect. By examining the specimen can we answer such questions as, when, where and why was this glass made, what was the source of its raw materials, what is it, or what is it a part of? Can it tell us what happened to the site where it was found? These questions have been answered in part in the preceding chapters, but in the context of the scientific and historical aspects of glass. In order to put glass in its proper perspective, and to provide background information in a logical way it is necessary to present the material in this manner, but I shall now turn the subject around and see if we can answer some of the important questions that the archaeologist will have in mind on finding a piece of glass.

At this point it is appropriate to give a word of warning. I have mentioned in other places that glass is one of the most complex of substances, especially ancient glass with its huge variety of raw materials, methods of manufacture and subsequent treatment. Its scientific study, as a disordered, multi-component system, is in many ways still in its infancy, though it has been studied intensively for over 60 years, and scientific interest has been shown for over 200 years. Although great progress has been made the difficulties encountered in development work with glass which involves a significant amount of scientific study of its nature and properties have often been considerable: the development of the float process by Pilkington Brothers for producing flat glass, now used world-wide, took over seven years of intensive work before it became a commercial proposition. At the present time the production of a glass suitable for storing radioactive waste is causing even more problems. Prediction based on scientific evidence is more difficult for glass than for most other materials and must be backed up by extensive experimental work. It follows that conclusions that can be drawn from examination of glass have to be very circumspect and are limited in scope. The amount of information that can be obtained can be increased by using all the scientific methods of investigation at our command: indeed, this is essential, but it would still be misleading to pretend that this will give us all the answers because there are so many gaps in our understanding of the simplest systems. I make no apology for repeating this point: in my opinion a great deal of harm has been done to the study and understanding of ancient glasses by people who built up great towers of speculative theories based upon the slenderest of scientific evidence. Fortunately there has always been a small body of people interested in scientific, artistic and social aspects who have realized the immensity of the task and I believe that this number is now increasing.

Material from glassmaking sites

Having presented what some may think a rather gloomy view, I should say that it is quite possible to draw valid conclusions, or, to be more accurate, to consider a range of options when looking at a piece of glass. First, consider where it is found. Is it a site where glass was made, or re-worked, or simply used (as a window material, container or for some other more specialized function). Glasshouses have been discussed in Chapter 6, and for later sites it is often possible to build up a picture from contemporary written records, extensive excavation, etc., but care should be taken in interpreting finds. The reason for this lies in the nature of the glassmaking process. In addition to the basic raw materials (sand, soda-ash and limestone for a typical modern commercial glass), waste glass of the same type, known as cullet, is added to the glass, because it melts at a lower temperature than any of its separate constituents: this speeds up the melting process and cullet can form a significant proportion of the total batch. There is also the advantage that wastage can be cut down by re-cycling unacceptable glass. The addition of cullet has a long history and you may expect to find samples at excavated glasshouses. This can be misleading, as the cullet could have been purchased from elsewhere and so could give a false picture of the products of the glassworks. Nowadays the composition of the cullet is most carefully matched to the basic glass composition but in the past, without benefit of scientific analysis, the situation was not so well controlled. Batch recipes show an amazing variety of constituents: for example, the following recipe for black bottle glass of the mid-nineteenth century reveals a decidedly casual approach: "sand, 3 barrows; lime, 4 barrows; red clay, 4 barrows; rock salt, 60 lb.; soap waste, 28 lb." However, chemical analysis and comparison of glass remaining in crucibles and supposed glass products can often identify and confirm the particular items made at the glassworks, although the situation is more complex if a range of types was produced.

Obviously, a complete object usually yields more information than a fragment of glass. Identification of both demands a background knowledge of glass history and design, and a certain amount of scepticism. For example, a particular colour of glass found on a site may not represent its usual output as it could have been discarded after an unsuccessful melting: changes in glass composition, furnace temperature and atmosphere, and melting regime can drastically alter the colour of the final product. As with all glass remains, weathering taking place since the glass was made will also affect the appearance of the specimen.

The further back one goes in time the more difficult it often becomes to gain clues from the site of glassmelting activities, because of lack of supporting evidence. The enterprise involves an increasing amount of detective work. Three examples which illustrate this point, and also provide a demonstration of what can be deduced from simple observations combined with more complex scientific analysis are the cases of the Beth She'arim glass slab, the Tell-el-Amarna crucibles, and the Celtic vitrified forts. These studies are described below in their archaeological contexts because the points that they illustrate are of less value if pulled out and quoted in isolation: indeed, any general statements made along these lines can be very misleading and it is best to study actual cases which suggest techniques applicable to your own investigations.

The great glass slab of Beth She'arim

The story of the investigation of the Beth She'arim glass is given by Brill (1967). Dr. Brill was a member of an American archaeological expedition who worked on the problem as one of their projects in the summers of 1964–6. Beth She'arim is an ancient site in southwestern Galilee, in a region where glass was produced during the Roman period, and probably both before and after this time. The site was formerly a famous centre of Jewish learning, with catacombs yielding important artistic finds. In 1956 an ancient cistern adjacent to the catacombs was cleared out with the intention of making it into a small museum for display of objects found in the catacombs. During the clearing operation the bulldozer ran up against a large rectangular slab: as this slab was too large to be moved it was left in position and the surrounding floor was levelled even with its base.

The suggestion that the slab might be made of glass was put forward by the members of the American expedition of 1963. During the following summers three small-scale excavations were made around the slab and laboratory studies of samples taken from it were also carried out. These established that the slab was made of glass and had been produced intentionally, not as a natural geological material. The chemical composition of the glass was consistent with compositions of Roman period glass, and although the lime content was about twice that usually found in glasses of the period, several other similar glasses with high lime content have since been found in the region, thus linking the slab to the local production of glass. The intentional nature of the melting was revealed when the team excavated directly beneath the glass and found large, well shaped limestone blocks in contact with the base of the slab. These must have formed the floor of the huge container in which the glass was melted.

The size of this early "glass tank" is astonishing, because the slab measures about 3·4 m × 1·95 m × 0·5 m and weighs about 8·8 tons, over half the weight of the 200 inch reflecting mirror in the Hale telescope at Mount Palomar. The usual process of melting at the time involved the use of small crucibles or pots, and it was not until the nineteenth century that melting in a large rectangular tank was successfully developed. The Beth She'arim furnace could perhaps be seen as a very early forerunner of the modern tank furnace.

It was also possible to deduce how, and under what conditions the glass had been melted by examining core samples taken through the slab. One reason why the fact that the slab was made of glass was not realized earlier was that the surface was weathered and opaque: only fracture surfaces revealed the typical concoidal fracture pattern of glass. When the core samples were sawn vertically, ground and polished, continuous cross sections through the total thickness of the slab showed that the glass was as much as 60% crystallized in many places, causing the opacity. The crystals were formed because the glass cooled down too slowly through the range of temperature just below the temperatures at which the materials were melted: the atoms in the glass thus had time to rearrange themselves in the crystalline configuration before the material reached a temperature at which its viscosity was too high for the rearrangements to occur (see Chapter 1). As well as observing this fact it was also possible to draw other conclusions. The core samples showed a top band of fairly uniform glass, blending into a frothy white band, in turn blending into a buff coloured band at the bottom of the slab which had a gritty texture. As analyses gave identical chemical compositions for all three bands, the bottom two bands represented original batch materials which had not been heated sufficiently to melt them completely. The batch materials least affected by heat were in the bottom layer: therefore heat must have been applied from above rather than below. This implies some type of reverberatory furnace, with the heat from the furnace fires being reflected down from a roof onto the glass tank below.

Brill then considered related historical factors in order to determine a date for the construction of the tank. Beth She'arim flourished between the second and fourth centuries AD, but at this stage the area where the slab was found had considerable religious significance, and it seems improbable that a glass furnace would be built there. However the city was destroyed following the crushing of a Jewish rebellion in 352 AD and it is unlikely that glass manufacture on a large scale could have flourished for very long under such conditions. Brill therefore considers that the evidence points to a date between the fourth and early seventh centuries AD, after which period site occupation is not consistent with the production of such a slab.

The Beth She'arim study is an excellent example of how deductions can be made about glass compositions, methods and date of manufacture, furnace construction, etc. Unfortunately we are still not sure why the slab was made. It seems that glass may have been melted at Beth She'arim in the tank and then broken up and sent elsewhere to be remelted and fashioned into objects. This double process was common, and was favoured because in many cases it would have been much more difficult to start with the raw materials of glass than to reheat and shape an existing glass. However, the poor quality of the glass in the slab, of uneven composition and showing a marked tendency towards crystallization, would have provided a poor starting material for the manufacture of glass objects. Also the glass was coloured purple, by the deliberate addition of manganese, which suggests that it could have been made as one special piece of glass, perhaps for architectural or monumental purposes.

Glass manufacture in ancient Egypt

Beth She'arim was concerned with glass melting on a large scale and was therefore rather unusual for the period in question. A study was made by Turner (1954) of the small crucibles employed in ancient Egypt for glass manufacture. The use of such crucibles was widespread but little was known about their chemical composition and physical characteristics, which in turn could provide information for archaeologists on melting temperatures and processes in use in ancient times.

The crucibles studied were originally assembled following excavations by W. M. Flinders Petrie in 1891 and 1892 at Tell-el-Amarna on the east bank of the Nile. This was the site of the new capital founded in the early fourteenth century BC by King Akenhaten: it was a place of great splendour and a centre of craftsmanship. Amongst the craftsmen were makers of glass and glazes. Numerous pieces of broken glass vessels, glass rod and fragments were found, together with a small number of crucibles.

Flinders Petrie distinguished three types of vessel used for melting glass. The first was a shallow, saucer-shaped pan, ten inches across and three inches deep in which material was fritted: this process, which involved preliminary heating of the raw materials prior to melting proper at higher temperatures, was common in the past. It assisted in the initial stages of melting (and also burnt off impurities) at a time when high temperature furnaces were not available. The pans contained a mass of vesicular, partly fused frit. The second type of vessel was a cylindrical pot, seven inches in diameter and five inches deep, externally. The third type was a smaller pot which Flinders Petrie considered may have been used for melting frit into glass: no remains were found but the size and shape were deduced from the outline of pieces of glass which appeared to have been cooled down from the molten condition in a refractory vessel.

Turner performed his analyses on the cylindrical pots. The type was easy to characterize because a number of fragments of varying size were available. They could have been made near Qena, some 80 miles south of Tell-el-Amarna, where a pot-making industry of long standing still existed. The pots were symmetrical, having been made on a wheel, and appeared to have been fashioned to a standard size as several fragments had similar dimensions. The question then arose as to what the cylindrical pots had been used for. Some were certainly for glassmaking, because blue glass was found adhering to the inner walls or bottoms of two pot fragments. Some pots had not been used but a third type of fragment was especially interesting. In several cases broad stripes of glass were discovered running from the bottom of the pot towards the rim. This means that the pots must have been in an inverted position when an overflow of glass occurred, running down the outside of the pot. Flinders Petrie had suggested that inverted cylindrical pots had been used as supports for the small pans in which the material was fritted. Any overflow from the pans would run down the sides of the pots in the way that was observed. Thus the pots appeared to have at least two uses.

When Turner analysed the pot material he found a complex composition with very high contents of iron oxide, calcium oxide, magnesium oxide and alkalis (sodium and potassium oxides): these would all have been detrimental to the refractoriness and stability of the pots. In order to find out what temperatures the pot would withstand, small pieces were heated for one hour at temperatures of 1100°C and higher. The results showed that the pot material would become fluid when subjected to prolonged heating at 1150–1200°C, and in the presence of a glass with high alkali content the breakdown temperature would probably be below 1100°C. Thus glass could not have been melted at higher temperatures, and the question arose as to whether a satisfactory material could be produced (by comparison, modern glass melting furnaces typically operate at about 1550°C). Turner therefore analysed the blue glass adhering to some of the pot fragments and prepared a batch of raw materials which would give a glass of the same composition. He ran a series of trial melts and found that it was possible to melt a good, clear, transparent glass, not completely free from bubbles, but the process took about 16 hours (although preliminary fritting would have reduced this time). Turner concluded that the upper temperature limit in glass melting operations of the period did not exceed 1100°C, and that the numerous bubbles in ancient glasses arose from the inability of the glassmaker to apply a high enough temperature to eliminate them, necessitating a prolonged heating period to minimize this problem.

Celtic vitrified forts

Vitrified forts are a group of prehistoric fortifications which were probably built during the early Iron Age, located on hilltops in Germany, France and Great Britain, especially in the Scottish Highlands. They have been studied for over 200 years, and many conflicting theories for their origin and purposes have been put forward. A recent review article which summarizes the earlier work has been written by Youngblood *et al.* (1978): these authors have also carried out an extensive chemical and petrological study which is described below.

The remaining walls of the forts are composed of cobbles fused together into breccias by a glassy, vesicular matrix which has flowed around the constituent rocks, welding the whole into masses that are several cubic metres in volume. Some of the rocks show signs of internal melting with individual mineral grains surrounded by a flowing glassy matrix. Thus considerable heat, applied over an extended period of time, must have been applied, and the question is, firstly, whether this application was constructive or destructive, and secondly, what methods (in the widest sense) were used to produce the vitrified forts. If the vitrification was deliberate, it is reasonable to assume that easily fusible materials would have been selected, and a flux of some sort would have been used to lower the melting temperature. Although there was some evidence of foreign materials such as shells, bones, seaweed and wood ash, the general opinion at the time of Youngblood's study was that fluxing agents had not been used. Nor did field evidence suggest that rocks with special properties had been chosen for wall construction: the location of the forts appeared to have been controlled by topography and in most cases there was no evidence for transport of wall materials. In places where rocks might have been brought to the site from a distance, it seemed that easily fusible materials were not selectively chosen. Vitrification therefore appeared to be destructive in nature. A widely held view was that it was produced by the burning of timber-laced walls. Evidence for this is based on such factors as wall geometry and patterns of wood casts in the glassy materials. However, there is little supporting evidence from excavations as to how the walls were constructed, and other suggestions are of an even more speculative nature. Youngblood and his colleagues therefore did not seek to answer all the questions associated with the forts but to carry out the first comprehensive and systematic comparison of the chemistry of rocks and vitrified materials in order to determine whether, and to what extent a "normal" process of partial melting had occurred. By partial melting is meant that a system is heated above its solidus temperature (the temperature at which melting first starts to occur) so that a liquid phase is formed. (As the process of melting is complex, occurring over a range of temperatures, the rock would have to be heated further before the liquidus temperature, the temperature at which it would be entirely molten, would be reached.)

Eleven forts were chosen for investigation (one German, three French and the rest Scottish). The degree of vitrification was found to vary considerably from fort to fort, and also along a single wall within a given fort, implying differences in heat intensity and materials being melted. The composition of the glasses was very heterogeneous: this would result from *in situ* melting and lack of convective mixing, revealed under the microscope by the range of light and dark colourations observed. Residual quartz grains were noticed, often in a highly fractured state which implied heating at a temperature of over 1000°C, and also small metallic spherules, consisting mainly of iron, which indicated strongly reducing conditions during melting. The use of timbers as part of the wall structure was deduced from the frequent presence of wood casts in the surface textures. Youngblood and his colleagues then carried out detailed analyses of composition of rocks and glass samples using the electron microprobe (see Chapter 3). The results, together with an analysis of the geochemical factors involved, allowed them to deduce that the observed chemical behaviour could be explained in terms of partial melting, *in situ*, and without the addition of fluxing agents or the use of more fusible rocks, thus supporting previous hypotheses based upon archaeological and other evidence.

High phosphorus and calcium contents in some of the glasses, together with surface textures which could be bone casts, suggest that in certain cases bone debris was used to fill the spaces between the rocks, giving a densely packed structure. Such compacted walls would have a good chance of withstanding a fire and might even have been strengthened by it. A further wet chemical analysis of the ferrous/ferric oxide ratios in the rocks and glasses showed that they were in all cases higher in the glasses than in the rocks, and thus that the environment during melting was strongly reducing. This in turn suggests that the fire was not open but in some way enclosed. Under these conditions combustion of timber present as part of the wall structure would proceed as in a charcoal kiln: burning would be slow and gases would be produced which would burn at even higher temperatures. This model provides a plausible mechanism for the attainment of the high temperatures necessary to cause partial melting of the rocks. It thus appears that the walls were built in such a way that if they were set on fire by an enemy they could withstand the heat and might even be strengthened by it.

Glass finds from archaeological sites

The previous sections have dealt with deductions that can be made about material found on a wide variety of glass-manufacturing sites. I hope that the importance of reaching such deductions in the context of the whole body of archaeological, scientific and other evidence has been made clear. Such considerations also apply to the study of glass finds at other archaeological sites. You should always look at the object carefully and adopt a critical approach to what you see. As explained in the chapter on conservation, variations in durability of various types of glass may result in only the most stable still remaining to be found, and this can distort the picture that you form of the original production. Weathering crusts and other forms of surface attack can greatly alter the appearance and dimensions of the object: for example a very fragile glass may not have been so originally because the surface layers have flaked off, and what appears to be metallic decoration is often caused by interference colours in thin weathering crusts. Also a modern piece of poorly durable glass can often be mistaken by the inexperienced for something considerably older. This problem is increased by the fact that, as explained in the sections on dating in Chapter 3, no method exists for the scientific estimation of age (except in a few very specialized cases).

The best way to become familiar with the appearance of the type of glass that you are likely to come across is to study actual examples. In this way you can learn to recognize not only the different styles and periods but to distinguish between glasses by observing the differences in quality and tint of the actual material. The "colourless" glass of the eighteenth century, used to make the beautiful drinking glasses of the period, lacks the water-like clarity of modern crystal glass, having by comparison a greyish tinge. A description of major collections of all types of glass in Great Britain is given by Ruth Hurst Vose (1980) but mention may be made here of the large and representative collection in the British Museum and the excellent Pilkington Glass Museum at St. Helens which covers all aspects of glass. As well as a fine selection of artistic glass the scientific and technical developments of the past 200 years are treated in depth and archaeological evidence is represented by material excavated from glasshouses and furnace reconstructions. This is probably one of the most interesting places for the archaeologist to visit, and the staff have been involved over the years with several glasshouse excavations.

You should also read up about the subject, and suitable books on various aspects of glass are suggested in other chapters. Finally you should understand the processes by which glass was made, as this will help you to interpret any finds that you may make. The techniques of glassmaking have been described briefly in Chapter 2 but the following sections will deal with two major areas of manufacture, containers and flat glass, in more detail and also draw attention to points that you can look for in examining these types of glass.

Containers

Although glass has been formed into hollow objects for over 3500 years, the art of shaping glass on the end of a blowpipe was not developed until much later, probably about 40–50 BC. As described in Chapter 2, the earliest vessels were frequently made by the slow and laborious method of wrapping glass around a core, perhaps made of mud bound with straw, which was then scraped or chipped out. Alternatively, the core, on the end of a metal rod, was dipped into the molten glass to collect sufficient material to make the neck and body of the vessel. Often the inside of such a vessel will show signs of having been in contact with the core, and there may be small pieces still adhering to it. The surface of the vessel was frequently decorated with coloured blobs or trails, which were pressed into the surface by marvering (rolling on a flat stone slab). These trails were commonly combed to make a festoon or zigzag pattern. A comprehensive account of these vessels, together with those made by other techniques in the pre-Roman period is given in the book *Masterpieces of Glass* compiled by Harden *et al.* (1968): this work also contains many pictures showing the great variety of types that were produced.

When glassblowing was first introduced the vessels were free-blown, but later two- or three-section moulds of wood or clay were used. On such glass, as on modern containers, seams, raised lines where the sections of the mould came together, are visible. For the first time it was possible to produce thin, transparent vessels on a much larger scale than had been possible by the older forming methods. The difference that this made to glassmaking may be judged by comparing the core-formed jug dating from the fifth century BC shown in Fig. 22 and the beautiful Roman blown flagon, Fig. 23. The new glassblowing technique was used to make containers for a wide range of products. Food could be stored in jars with wide, flat bottoms which stood upright as opposed to many Egyptian containers which, formed round a core, were narrow-based and required a stand to hold them: wines, oils and medicines could also be conveniently stored and transported. Glassmakers were proud of their wares and often marked their moulds with their trade names, for example, Jason, Artas and Ennion. The new range of products accentuated the problems of sealing the vessels and during the Roman era sealing techniques included the use of cloth or cloth soaked in wax or oils, fibres soaked in oil, and grease or wax gobs. Linen cloth treated with oils or resins has a very long life and was used as a "tie-on" cover from earliest times for sealing vessels which contained both liquid and dry materials.

After the fall of the Roman Empire the use of glass declined and styles became simpler: a typical example of a bottle made during this period is shown in Fig. 24. The plainer glass may not reflect a loss of glassmaking skills but rather the different tastes of the Teutonic patrons. The gradual revival of the art of glassmaking in the Frankish Empire does not appear to have greatly affected the manufacture of containers, and only small bottles, such as those used by doctors and alchemists, appear in significant quantities throughout the Middle Ages. Indeed, the method of manufacture of bottles altered little for nearly 2000 years. In the early nineteenth century bottles were still blown either off-hand without a mould or in an open single piece mould which only shaped the bottom half or body of the bottle. The upper or shoulder parts together with the finish (the top of the neck) were shaped while still soft by applying a tool as the bottle was rotated. It is interesting to note that the word "finish" is still used though this is now the first part of the bottle to be formed on modern container machines.

The use of glass bottles and containers for a variety of purposes began to grow again during the seventeenth century, with a consequent increase in the variety of vessels that can be found during archaeological investigations. The following sections deal with the major types and their development.

Wine and beer bottles. Until the beginning of the seventeenth century nearly all bottles were made of earthenware, leather, metal or wood, none of them an ideal material for containing wine. The growth in technical knowledge in the glassmaking industry at this time, which

Fig. 22. Oinochoe, fifth century BC, from Camiros, Rhodes, made by dipping a core into molten glass or winding threads of glass around it. Decorative threads were trailed around the jug and pressed into the surface by rolling. The handle and foot were added and the core was then chipped out. Reproduced by courtesy of the Trustees of the British Museum.

Fig. 23. Roman blown glass flagon, late second or early third century AD, from Bayford, Kent. The production of this thin, transparent and elegant vessel was made possible by the invention of glass blowing. Reproduced by courtesy of the Trustees of the British Museum.

Fig. 24. A free-blown "pouch" bottle, seventh century AD. Styles became much simpler after the fall of the Roman Empire and many of the Roman techniques fell into disuse. Reproduced by courtesy of the Trustees of the British Museum.

may be said to have started with the publication, in 1612, of Antonio Neri's *L'Arte Vetraria*, was rapid, spreading from Italy to France and England where there was an upsurge in bottle-making. Glassmakers had available to them the 1662 translation of Neri's book by Christopher Merrett. Dr. Merrett, a founder member of the Royal Society, added his own extensive "Observations", doubling the size of the original work. The times were also propitious, there being a great increase in wine drinking following the restoration of Charles II in 1660. Samuel Pepys tells of a lady, Mrs Shippman, who filled a vessel "full of white wine, it holding at least a pint and a half, and did drink it off for a health, it being the greatest draught that ever I did see a woman drink in my life."

The increase in the use of bottles was also brought about by improvements in closures. Cork was increasingly used during the seventeenth century, but until the introduction of the corkscrew at some time prior to 1686 it was not possible to use tightly fitting corks, and the stoppers had to be tied on with string or wire. This may seem an unimportant invention but it led to horizontal storage, maturing of wine in bottles and the possibility of producing sparkling wines by the *methode champenoise*. Tied stoppers required an anchorage, and the earlier bottles had a definite rim, the string-rim, around the bottle just below the mouth. The shape of the bottle continued to change throughout the seventeenth and eighteenth centuries, thus affording a valuable method of dating bottles found at an excavation site. The earliest wine containers were bulbous with a round base, pale green in colour and very light in weight. They were kept upright by special metal table stands or wanded by encasing in osier basketwork woven at the glassworks: some Italian wines are still bottled in a similar way. Bottles manufactured during the first half of the seventeenth century are called "shaft-and-globe" being nearly globular in body with a long slender neck. This shape was easy to blow but difficult to balance and in the third quarter of the seventeenth century the body became squatter, the neck shorter and the base flattened to provide more stable support. The kick-up or basal concavity was increased in size: many explanations have been offered for this but it did improve stability. These kick-ups, which seem to become easily detached when a bottle breaks, can cause confusion when found on a site because of their considerable size: it is helpful here to examine complete specimens to see how the bottles were formed. Harvey's Wine Museum in Bristol has some interesting examples, as well as a superb collection of English drinking glasses and decanters.

Although the appearance of the squat bottles is attractive they were difficult to store and a taller, narrower bottle with straight, slightly slanting sides was developed after about 1715. By 1750 the slanting sides had become vertical, and a cylindrical bottle, very similar to the modern port bottle was in use. Figure 25 shows the developoment of the wine bottle.

It is often possible to date a bottle because it carries an embossed seal. These were affixed to bottles sold to clubs, inns and the more well-off section of the community. Some carry the date and the initials or crest of the buyer and were sealed to all bottles intended for the client. Wines and spirits were stored by the inns and the customers had their bottles filled from these supplies.

Like wine, the origins of beer go back before the beginning of recorded history but again, until the seventeenth century it was stored not in glass but in leather containers. As the century progressed home-brewed ale was stored in bottles which were made either of earthenware or glass and the shape of the glass beer bottle evolved in a similar way to that of the wine bottle. During the late eighteenth and nineteenth centuries, prior to the introduction of automatic bottle making, a great variety of bottle shapes were introduced, some of them poorly adapted to the blowing process. Bottles intended merely for containers were black or very dark green through excess iron and other impurities in the raw materials. In 1831, in evidence given at an excise enquiry, it was stated that the materials employed in common bottle manufacture were sand, soap-makers' waste (containing soda), lime, common clay and ground bricks. It is no wonder that finds can be misleading with these widely differing compositions, owing to variations in stability of the different glasses. Quite modern glass can have a spurious appearance of age, having undergone considerable attack owing to an unstable composition. As pointed out in Chapter 3 it is not possible, except in very special cases, to determine age scientifically, and so a knowledge of styles and the history of glassmaking must be combined with other evidence from the site, study of local archives, etc., before you can come to any valid conclusions.

Bottles for soft drinks. The growth of this type of bottle occurred somewhat later than that of beer and wine bottles, though a high proportion of the containers found by industrial archaeologists interested in more recent history are of this type. The drinking of natural spa water became popular amongst the English upper classes in the eighteenth century, but this water often had a very unpleasant taste and although it

Fig. 25. The development of the wine bottle. The first wine bottles were virtually blown bulbs of glass but as techniques improved they assumed a cylindrical shape and by 1750 a bottle very like the modern port bottle was in use. Reproduced by courtesy of the Glass Manufacturers' Federation.

was sold, in earthenware containers, for its (supposedly) curative properties it did not have a wide market. Attempts to produce substitutes date back to the mid sixteenth century when it was recognized that "gas" was present in the water, but it was Joseph Priestley in the 1770s who succeeded in developing the first practical method of making artificial mineral water, containing carbon dioxide. Commercial production of soda-water began in Manchester in 1777, and after the invention of the soda-water syphon in 1815 it rapidly became a popular bottled product, over one million bottles of flavoured mineral water being sold at the Great Exhibition of 1851.

Glass bottles were very quickly used for mineral water as the early bottles, made of earthenware, were permeable at high gaseous pressures. In 1814 William Hamilton patented an egg-shaped bottle for artificial mineral water. Because of its shape it had a much greater resistance to high internal pressures than the bottles in general use at the time, and as it had to be stored on its side the cork was kept moist, thus preventing leakage of carbon dioxide through a dry stopper. This bottle was widely used in England after 1840 until the end of the century, when it was replaced by the flat-egg bottle which was easier to fill and could be stored on its side or on its flat narrow base. It was adapted to take the crown cap in about 1903 and remained popular until the late 1920s. The famous Codd bottle was patented by Hiram Codd of Camberwell in 1875. It contained a glass marble which was kept pressed against a rubber ring in the neck of the bottle by the internal gas pressure, thus forming an excellent hermetic seal which was released when the marble was forced downwards. It was very popular in Britain from 1890 to 1914, though it continued to be used here until the 1930s and is still made in the Far East. A variety of soft drink bottles are shown in Fig. 26: they illustrate some of the main types that you may come across when digging in glass manufacturers' waste tips or town dumps. Such dumps are often opened up by excavations for new roads or buildings and provide an excellent source of information for the archaeologist as well as the bottle collector.

The best way to become acquainted with bottles in all their variety is to look at actual examples, well documented, in collections, such as Harvey's Wine Museum, mentioned above. For those who cannot visit such collections there are an increasing number of books on the subject of containers. Coloured pictures are especially useful in this respect, and the book by Fletcher (1976) is a good example. For those who want to investigate the history and development in more detail,

Fig. 26. Nineteenth century soft drink bottles. In the foreground is the Hamilton bottle; in a semi-circle from left to right, a tall narrow Seltzer bottle, a "flat-egg" bottle, a ball-stoppered bottle, and another Seltzer bottle. Reproduced by courtesy of the Glass Manufacturers' Federation.

Meigh (1972) gives an authoritative account in his *Story of the glass bottle*. Another useful book by Wyatt (1966) surveys the whole history of glass containers, and Talbot (1974) has produced a review article on the evolution of the glass bottle for carbonated drinks, with many illustrations.

Glass for medicine, perfumes and cosmetics. The use of containers for medical and related products also increased greatly during the late eighteenth and nineteenth centuries (although the use of "urynalls", medical inspection bottles for the examination of urine from patients was common throughout the Middle Ages). Patent medicines were becoming popular, and one could buy anything from hair restorer to stomach medicine from vendors who travelled from town to town or displayed their wares at country fairs. Glass was also used for pharmaceutical purposes. Daniel Defoe, in 1727, stated that "fine flint glass (including) apothecaries and chymists glass phyals, retorts, etc. (are made at) London, Bristol, Sturbridge, Nottingham, Sheffield (and) Newcastle." The development of pharmacy in England was complex, with several different groups vying with one another to prepare and sell medicines and offer medical treatment to the patient, and until 1868 anyone could open a shop as a chemist and druggist. A great variety of containers appeared, often being used for colourful displays in shop windows to tempt prospective customers. These displays became very popular in the late eighteenth century, coinciding with the introduction of new shop fronts with larger panes of glass which allowed containers to be shown to the best advantage.

The history of pharmaceutical glassware is described in detail in the book by Crellin and Scott (1972) which covers the period from 1600 to 1900 drawing on examples from the Wellcome Collection of British Glass. As well as containing a large number of pictures of this outstanding collection there is an extensive bibliography for further reading.

Flat glass

The history of glass as a glazing material has been given briefly in Chapter 2. In some ways it is more difficult to draw conclusions from finds of window glass, especially small, shattered pieces, than it is from vessel glass. Prior to the introduction of mechanized processes most glass was either of the crown or broad glass type (although high quality cast plate for the luxury market was introduced in the late seventeenth century). The processes changed little over the centuries so it is not generally possible to come to conclusions about age by examining pieces of window glass. Broad glass was made from a blown cylinder (Fig. 6, Chapter 2) and so bubbles present may tend to be elongated along the axis of the cylinder, though this is not conclusive evidence that the piece was made by this process. Also, during manufacture the cylinder was cut and flattened to form a sheet: thus the surface came into contact with the flattening tool and so lost its natural fire polish. Crown glass, on the other hand, did not come into contact with anything that would dull its surface and its natural polish theoretically forms a basis for distinguishing between the two types. Unfortunately, over the years, especially for buried glass, this polish may well have disappeared owing to glass attack. However, crown glass often shows spin marks, concentric circles, which developed as the glass was spun rapidly to open it out to form a disc on the end of the metal rod. The "bullseye", the thickened knob of glass at the centre of the disc consists of poor quality glass, uneven and often cracked, but it was better than no glass at all and can often be found in the windows of humbler domestic buildings.

Stained glass owes its colours to metal compounds added to the batch to give a body colour, or applied as a pigment to give surface decoration (see Chapter 2). A deep green colour in window glass is often mistaken for a sign of age: as explained before a very small amount of iron, present as an impurity, causes this intense colour, and the problem was not overcome until recent times when raw materials of high purity and constant composition became widely available. Also look out for the "flashed" layer of copper ruby glass applied to a base of colourless glass by dipping in the melt because of the intense colouring effect of copper oxide. One colour of glass which may be misleading is purple: glasses to which manganese was originally added as a decolourizer may, through long exposure to sunlight, have developed this colour, because the light has effectively oxidized the manganese back to the purple form.

Bibliography

Brill, R. H. (1967). A great glass slab from ancient Galilee. *Archaeology* **20**, 88–95.

Crellin, J. K., and Scott, J. R. (1972). *Glass and British pharmacy 1600–1900: a survey, and guide to the Wellcome Collection of British Glass.* (Wellcome Institute of the History of Medicine)

Fletcher, E. (1976). *Antique bottles in colour.* (Blandford)

Harden, D. B., *et al.* (comps.) (1968). *Masterpieces of glass.* (Trustees of the British Museum)

Meigh, E. (1972). *The story of the glass bottle.* (C. E. Ramsden & Co.)

Talbot, O. (1974). The evolution of glass bottles for carbonated drinks. *Post-Medieval Archaeology* **8**, 29–62.

Turner, W. E. S. (1954). Studies of ancient glass and glass-making processes. Part 1. Crucibles and melting temperatures employed in ancient Egypt at about 1370 BC. *Transactions of the Society of Glass Technology* **38**, 436–444T.

Vose, R. H. (1980). *Glass.* (Collins)

Wyatt, V. (1966). *From sand-core to automation: a history of glass containers.* (Glass Manufacturers' Federation)

Youngblood, E., *et al.* (1978). Celtic vitrified forts: implications of a chemical-petrological study of glasses and source rocks. *Journal of Archaeological Science* **5**, 99–121.

8

Searching for Information on Glass

In the previous chapters I have given references to key articles and recommendations for further reading. Some of these references will retain their value, but in all fields of interest to the archaeologist the picture is constantly changing. Therefore it is of prime importance to be able to keep up with the latest publications, whatever the area of research.

The aim of this chapter is to enable you to search for your own information on glass and to be reasonably certain that you are covering your area of interest in a systematic and comprehensive way. In order to do this you need to use secondary guides to the primary research literature, as it is clearly impossible to identify, let alone have the time and the ability to search through all the original material that may be of interest to you.

In this chapter I have made a selection of sources from amongst the much larger number that might be of interest to a specialist in a particular area of glass research. The pattern of searching that I advise is also my own, one that I have found to be most useful in my research, but you should feel free to vary this according to your own requirements. Most of the searching tools that are mentioned are available in a large academic or public library. If you do not work in a university department, it is usually possible to obtain permission to consult the material in your nearest university library, if you have good reason for doing so.

The search sequence described in the following paragraphs can be entered at any stage, depending upon the amount of knowledge that you already have of the subject. However, even though you may be quite familiar with a particular aspect of the subject it sometimes happens that you have to research a topic that is entirely new to you, and I shall therefore start with the steps that you should take at the beginning of a search.

First it is necessary to find an introduction to the subject, in order to provide a framework for the gathering of more facts, and also to suggest subject keywords with which to search for relevant references. Depending upon your area of interest there is a wide variety of dictionaries and encyclopaedias. An excellent specialist dictionary is *An Illustrated Dictionary of Glass*, by H. Newman (Thames and Hudson, 1977). This includes over 2400 definitions of wares, materials, processes, forms and decorative styles, and entries on principal glassmakers, decorators and designers, from antiquity to the present. There is an introductory survey of the history of glassmaking. The work contains over 600 illustrations, successfully combining artistic and scientific aspects of glass. Entries are often detailed and are adequately cross-referenced. Important bibliographical sources are listed for further reading on particular subjects.

Another dictionary that is worth consulting is *The Collector's Dictionary of Glass*, by E. M. Elville (Country Life, 1961). Although intended primarily for collectors it provides introductions to many aspects of glass and glassware by means of extended entries. It is particularly valuable for its coverage of great British glassmakers and decorators.

The *Encyclopaedia Britannica* (Encyclopaedia Britannica, Inc., 15th ed. 1974) contains an extended entry on glass. Sections deal with the history of glass from antiquity to the present day; glass products and production; and a description of the glassy state. Written by experts in the various fields, there are extensive annotated bibliographies, although these would now need updating for the period after about 1970.

Another good source of introductory surveys is *A History of Technology*, edited by C. Singer *et al.* (Clarendon Press, 1954–78). This major comprehensive work covers, in seven volumes, the period from early times to the mid-twentieth century, and nearly all the volumes have separate entries on glass manufacture, written by specialists, with extensive annotated bibliographies. Although primarily a technological history, glassmaking is also placed in its economic and artistic contexts.

The next source of useful information is the published bibliography. The outstanding example is the *Bibliography of Glass*, by G. S. Duncan, edited by V. Dimbleby, subject index prepared by F. Newby (Dawson, for the Society of Glass Technology). This covers the period from the earliest records to 1940. Compiled and corrected over a period of 40 years, the *Bibliography* is a critical selection of 15 752 references, many with annotations, covering all aspects of glass. It is particularly rich in references to stained glass, to historical contributions, to records (including sales catalogues) of famous collections of glass and glassware, as well as to the production and working of glass. The entries are arranged by author, and there is a very good subject index.

It has proved impossible to extend the *Bibliography* for the years after 1940 in a similar manner and you must use other guides to the primary literature to cover this period. There are two useful bibliographies. The first is the *Isis Cumulative Bibliography: a bibliography of the history of science formed from Isis Critical Bibliographies 1–90. 1913–65*, edited by M. Whitrow (Mansell and The History of Science Society, 1971–). This guide has sections on personalities (with dates), institutions, and subjects. A further volume will be devoted to periods. The subject entries, some with short annotations, are arranged in a classified order, with a section being devoted to glass. There is also an alphabetical subject index. The compilation continues with the series of *Isis Critical Bibliographies*, which are issued annually as part of *Isis*, the official journal of The History of Science Society. Here the entries are arranged by period and broad subject division only, so that if you want to find all references on glass or related subjects, you will have to search in several subject sections.

Easier to search, and probably containing more relevant references is *Bibliographic Index* (H. W. Wilson Company, 1937–42—) which in fact is a bibliography of bibliographies. This is a subject list of bibliographies, in all subject areas, which have been published separately or have appeared as parts of books, pamphlets and periodicals. Selection is made from bibliographies which have 50 or more citations and so items found here will themselves provide useful reference lists. About 2400 publications are regularly examined, but the *Index* concentrates on titles in the Germanic and Romance languages. It appears three times a year, the final isue being the annual cumulation. Its advantage is that it covers a wide variety of material and thus picks up references outside the usual range of glass literature that might otherwise be difficult to trace.

Important information appears also in book form and here there are many finding aids relating to publications of particular countries, collections of national libraries, and books available for purchase. A range of these publications is stocked by most large academic and public libraries. The *British National Bibliography* (British Library Bibliographic Services Division, 1950–) is a catalogue, arranged in classified subject order, of books published in the United Kingdom. There are author/title and alphabetical subject indexes. To keep you up to date the *Bibliography* is issued weekly, but perhaps the most convenient publication is the annual cumulation. As all publishers are required by law to deposit copies of their books which then appear in the *Bibliography* it provides one of the best guides to books published on glass in this country. It goes back to 1950 enabling you to compile a list of books if you wish. Many of these will be out of print: books available for purchase are listed in *British Books in Print* (Whitaker). This is an author/title/subject listing, available as an annual printed volume. Many libraries now have a microfiche version, updated monthly, which will provide you with news of the latest books.

By the time that information has been published in book form it is already rather old and you will also need to keep abreast of the latest work in your field. If you had unlimited time available for searching through journals in the library you would still miss a great deal of important material, which could be out on loan or published in journals that the library does not take. It is a mistake to think that by regularly scanning a few well-known journals in your subject you will find everything that is relevant: you may well miss a key paper by following this routine and this is especially true of a subject like archaeology which overlaps with many other areas of study. You can cut down the amount of time that you spend, whilst at the same time doing an efficient and comprehensive search, by using abstracting and indexing publications. These secondary publications, covering hundreds or even thousands of primary journals in a particular subject area, provide access by author and by subject to the journal articles. Brief bibliographical details of the article are given and, in the case of an abstracting publication, a short summary.

The *Journal of Glass Studies* (Corning Museum of Glass, 1959–) is published yearly and contains a very extensive check list of recently published articles and books on glass, in its historical, economic and artistic aspects, from all areas of the world and periods. It also lists a few references of a technological nature. Abstracts are not provided. For those interested in these areas of glass studies this is probably the most useful source of information available, and the *Journal* itself contains authoritative articles on various areas of glass research.

To complement the *Journal of Glass Studies* the two journals of the Society of Glass Technology, *Physics and Chemistry of Glasses* and *Glass Technology* (1960–) contain large abstract sections and are the best sources of information on these aspects of the glass literature. They appear six times a year, with cumulated annual author and subject indexes, and have the advantage of providing detailed informative abstracts. *Glass Technology* in particular is useful as its abstracts contain a section devoted to art, design and history. Both journals also publish original articles, including papers on scientific, historical and economic aspects of glass studies of interest to the archaeologist. They continue the tradition of their predecessor, the *Journal of the Society of Glass Technology* which since 1917 had published important papers and critical abstracts on all aspects of glass studies. The journals also contain reviews of books on glass.

If you are concerned with any aspect of conservation of glass, the *CV News Letter* (*Comité Technique du Corpus Vitrearum*, 1972–) is an international publication devoted to the study and conservation of stained glass (although other types of glass are also considered). It contains news items of general interest, research in progress, articles on the history of glass, practical recommendations for conservation and requests for help or information. There is a section of book reviews and a bibliography, with detailed abstracts, dealing with important articles on conservation. At present the *News Letter* is published twice a year.

Two important foreign journals which contain abstract sections should also be mentioned, as they extend the range of journals covered by English language publications. *Verres et Réfractaires* (1946–) appears six times a year, and there are annual cumulative author and subject indexes to the abstracts, which are taken from a world-wide selection of journals. Abstracts include those on scientific investigation of ancient glass, and on art and history of glass. Each issue also contains an annotated bibliography of recent books.

The German publication *Glastechnische Berichte* (1923–) also contains abstracts from a wide range of journals, including a section on the history of glass. There is a certain overlap between the coverage of these journals, but you should consider consulting several publications in order not to miss important articles.

Specialist abstracting and indexing publications will provide you with much of the information that you will need, but it is worthwhile remembering that they cover only primary journals published in a narrow subject range. An important article may well have appeared in a journal which does not fall within this range. To cover this type of article adequately you need to use secondary publications in other subject areas. Some of the most useful are described below.

British Humanities Index (The Library Association, 1962–) is a subject listing of articles from about 270 British periodicals covering arts and social studies. It is issued quarterly with an annual cumulation (which also contains an author index). The *Index* is particularly useful for articles which have appeared in newspapers and in fine arts publications. Although these tend to be of a popular nature they provide introductions to areas of glass which tend to be neglected by research publications. *Art Index* (H. G. Wilson Co., 1929/32–) is a quarterly subject and author index, with annual cumulations, to art periodicals and museum bulletins. It covers all types of artistic glass and glassware, with sections on collections and exhibitions, and bibliographies of glass. *Artbibliographies Modern* (Clio Pr., 1969–) is a semi-annual abstracting publication which provides comprehensive bibliographic coverage of current articles from about 350 journals, books, dissertations, and exhibition catalogues on art and design. There is full coverage of 19th- and 20th-century studies, with articles on glass from all over the world.

On the scientific side the most useful general abstracting journal is *Chemical Abstracts* (American Chemical Society, 1907–). This is the world's most comprehensive abstracting service, covering over 14 000 of the most important scientific periodicals and other publications from all over the world. Because of this broad coverage it is invaluable for all scientific aspects of glass studies. Section 20, History, education and documentation, contains most material of specifically archaeological interest, but the detailed subject index provides access to articles on other aspects of glass, including the many studies on chemistry of ancient glass.

Another useful scientific source is *Engineering Index* (Engineering Index, Inc., 1884–). This covers over 3000 journals and in spite of its name is an abstracting rather than an indexing publication. It also contains references to a large number of reports and conference proceedings. There is an extensive section on glass covering such aspects as analysis, chemical attack, colour in glass, corrosion, microscopic examination and spectroscopic analysis.

The best source for traditional "archaeological" material on glass (reports on finds, etc.) is the *Journal of Glass Studies*, described above, but *British Archaeological Abstracts* (Council for British Archaeology, 1967–) supplements this information with articles (taken from about 230 publications) which relate in some way to archaeology in Great Britain and Ireland. Although glass is fortunate in having these two guides it is regrettable that in general the scientific areas are covered far more thoroughly, reflecting a general disparity between secondary publications in the sciences and humanities.

Not all the information of interest to you is contained in books, reports, conference proceedings and journal articles. Archival material, for example, whilst presenting special problems, can be a rich source of information. It is outside the scope of this book to guide you in this area, but many companies and public and academic libraries have their own archivist who will help you if you need to locate and consult original material. You should also consider patents, which can provide valuable background information on old processes and products. Collections of British patents, with associated finding aids, are located in several provincial public libraries, although the principal facilities in the United Kingdom for providing access to patent documents are found in London, at the Patent Office and the Science Reference Library, Holborn.

Finally, mention should be made of a specialist library in the field of glass studies, the Frank Wood Joint Library of Glass Technology (Department of Ceramics, Glasses and Polymers) at the University of Sheffield. This has an excellent collection of books, journals and other material of interest to archaeologists on the historic, artistic, scientific and economic aspects of glass, including interesting manuscript material and many out-of-print books.

Index